Design Calculations in Wastewater Treatment

Design Calculations in Wastewater Treatment

F. Wilson

Department of Civil Engineering,
University of Canterbury,
Christchurch, New Zealand

LONDON NEW YORK
E. & F. N. Spon

First published 1981
by E. & F. N. Spon Ltd
11 New Fetter Lane, London EC4P 4EE
Published in the USA
by E. & F. N. Spon
in association with Methuen, Inc.
733 Third Avenue, New York, NY 10017

Printed in Great Britain by
Richard Clay (The Chaucer Press) Ltd, Bungay

ISBN 0 419 11690 7 (cased)
ISBN 0 419 11700 8 (paperback)

British Library Cataloguing in Publication Data

Wilson, Francis
Design calculations in wastewater treatment.
1. Sewage – Purification – Mathematics
2. Sewage disposal – Mathematics
I. Title
628.3'01'51 TD743 80-41869

ISBN 0–419–11690–7
ISBN 0–419–11700–8 Pbk

Contents

Preface

The subject matter of this book arose out of courses in Public Health Engineering given by the author, at undergraduate, graduate and extension studies level in the UK at the University of Strathclyde, Glasgow and in New Zealand at the University of Canterbury, Christchurch. The text is set at undergraduate and introductory postgraduate level. It should also be useful to many engineers, both civil and mechanical, and to scientists working in the field of water pollution control who have not had intensive, formal lecture courses in public health engineering.

The format of the book comprises short theoretical sections followed immediately by a calculation designed to show the application of the theoretical relationships. The use of short sections, and therefore relatively uncomplicated calculations, is intended to give the reader confidence in selecting the correct design input data and in using the theoretical relationships within their context in order to achieve a realistic operational design. Where design relies heavily on experience as, for example, in the case of small scale sewage treatment processes, longer design calculations are presented. The derivation of theoretical concepts and the description of processes has been kept to a minimum since many excellent texts already cover these aspects; however sufficient theory and description is given for the book to provide a realistic introductory text. Numerical problems are given after the relevant chapters in order to help the student achieve confidence in process design calculations. The answers are given at the end of the book.

The text covers the water cycle from the precipitation and collection of stormwater, through the various conventional sewage treatment processes, to the effects of effluent discharge on the river system. In addition to the more usual large scale treatment processes, the text also covers the design of rotating disc units (both large and small scale), oxidation ponds (facultative and maturation ponds) and oxidation ditches. Calculations are also presented based on the newer pure oxygen and single vessel nitrification–denitrification systems.

Many engineering and science courses are so intensive that little time is available for students to learn information retrieval techniques. The author has therefore given suggested reading at the end of each chapter but has also given, in the Appendix, a guide to source information which should enable rapid updating of subject information.

The author wishes to thank his colleagues, including Professors R. Park and

I. R. Wood of the Department of Civil Engineering at the University of Canterbury, NZ for their enthusiasm. In particular a debt is acknowledged to Mr Ian Gunn of the Department of Civil Engineering, University of Auckland, NZ, who willingly accepted the task of checking the manuscript. I wish also to thank Dr D. Barnes of the School of Civil Engineering, University of NSW for his contributions to the nitrification–denitrification section of Chapter 5. Mrs V. Grey is thanked, not only for her excellent draughtmanship but also for her help and advice. Thanks are also owing to the many firms and authors who kindly gave permission to reproduce diagrams, tables, etc. Finally the author wishes to thank his wife for her patient help.

F. Wilson

Abbreviations and definitions used throughout the book

Abbreviations

BOD	biochemical oxygen demand
BOD_5	5 day biochemical oxygen demand
cap.	capita
COD	chemical oxygen demand
DO	dissolved oxygen
DWF	dry weather flow
g	grams
h	hour
l	litre
m	metre
MC	moisture content
mg/l	milligrams per litre
MLSS	mixed liquor suspended solids
s	second
SVI	sludge volume index
SS	suspended solids
TWL	top water level
30:20	30 mg/l SS and 20 mg/l BOD_5

Definitions

Activated sludge – A dark brown suspension of bacterial flocs produced by the aeration of sewage; prolonged aeration in admixture with this sludge effectively purifies sewage.

Baffle – A board which dips below the TWL of a tank at the inlet to disperse the flow into the tank.

Biological filter (also known as a bacteria bed or percolating filter) – A porous bed of suitable, graded, inert material. Bacteria and other organisms flourish on the surface of this material and bring about oxidation of the organic matter in the settled sewage applied to the filter.

Distributor – A device which distributes the tank effluent evenly over the surface of the filter medium.

Dosing chamber – A container which receives effluent from a septic tank or from a settlement tank, and from which this effluent is automatically discharged to the filter distributor, in intermittent doses.

Dry weather flow (DWF) –

(1) The sewage together with infiltration, if any, flowing in a sewer in dry weather.

(2) The rate of flow of sewage together with infiltration, if any, in a sewer in dry weather.

Filter medium – The material, such as hard clinker or broken stone, with which a biological filter is filled.

Final effluent – The liquid discharged finally from a sewage treatment works.

Five day biochemical oxygen demand (BOD_5) – The mass of oxygen consumed by unit volume of liquid (expressed as milligrams per litre (mg/l)) during biological oxidation in the course of 5 days at 20 °C.

Foul sewage – Any water contaminated by domestic waste or trade effluents.

Humus tank – A tank, through which filter effluent is passed, to settle solids which should be removed at frequent intervals.

Invert – The lowest point of the internal surface of a drain, sewer or channel at any cross-section.

Septic tank – A tank through which sewage is passed to settle solids which are retained to undergo digestion by anaerobic bacterial action.

Settlement tank – A tank through which sewage is passed to settle solids which should be removed at frequent intervals.

Sewage – The discharge from domestic and sanitary appliances on premises coming within the scope of this book.

Sludge – The slurry which is formed in a tank by deposition of settleable solids.

Storm sewage – Foul sewage diluted with surface water.

Storm overflow – A device, on a combined or partially separate sewerage system, introduced for the purpose of relieving the system of flows in excess of a selected rate, so that the size of the sewers downstream of the overflow can be kept within economical limits, the excess flow being discharged to a convenient watercourse.

Supernatant liquor – The layer of liquid overlying the settled solids which have separated from it.

Surface water (stormwater) – Natural water from the ground surface, paved areas and roofs.

Suspended solids (SS) – The solids which are suspended in a sewage or effluent.

Trade effluent – The fluid discharge, with or without matters in suspension, resulting wholly or in part from any manufacturing process and including farm and research institution effluents.

TWL – The top water level in a tank.

Water table – The surface of the subsoil water.

The majority of these definitions are an extract from CP 302:1972 and CP 2005:1968 and are reproduced by permission of the British Standards Institution, 2 Park Street, London W1A 2BS, from whom complete copies can be obtained.

1

Stormwater, sewers and sewage

Nomenclature

Symbol	*Definition and units*	*Dimensions*
w	Total rainfall over period t, mm	L
t	Storm duration in h (later reaction time in days for BOD work)	T
t_m	Storm duration, min	T
T	Number of storms of given intensity expected in 10 years as a long term average. Also the return period of the storm.	–
R	Intensity of rainfall w/t, mm/h	LT^{-1}
A	Catchment area, m^2	L^2
l	Sewer length, m	L
t_e	Time for surface water to enter main sewer, min	T
V_F	Velocity in sewer at full flow, m/s	LT^{-1}
t_c	Time of concentration of storm	T
Q	Flow in pipe or from catchment, m^3/s, l/h or l/s as stated	L^3T^{-1}
C	Impermeability coefficient (1 for impermeable surface) dimensionless	–
i	Average intensity of rainfall over time t_c, mm/h	LT^{-1}
A_c	Area contributing runoff, m^2	L^2
d	Pipe diameter, m or mm as stated	L
S	Gradient to which pipe is laid, m/m	–
n	Mannings coefficient of friction, $s/m^{1/3}$	$TL^{-1/3}$
A	Cross-sectional area of pipe taken up by liquid, m^2	L^2
P	Wetted perimeter of pipe, m	L
H	Fractional depth of flow	–

V_H	Velocity of flow at depth of flow $H \times d$, m/s	LT^{-1}
θ	The angle between line joining centre of invert and central point of pipe and the line joining centre of pipe and point where TWL touches pipe wall. See Fig. 1.2	radians
Q_H	Flow in pipe at depth of flow $H \times d$, m^3/s	L^3T^{-1}
A_F	Cross-sectional area of pipe, m^2	L^2
V	Velocity of pipe flow at unspecified depth of flow, m/s	LT^{-1}
Q_F	Flow in pipe when full, m^3/s	L^3T^{-1}
DWF	Dry weather flow. The average dry weather flow in the sewer l/d or m^3/d but also used as an instantaneous flow m^3/s or l/s (see definitions)	L^3T^{-1}
N	Population number served	–
G	Per capita daily water consumption, l/cap . day	L^3T^{-1}
I	Infiltration to sewer, l/day	L^3T^{-1}
T	Trade flow to sewer, 1/day	L^3T^{-1}
BOD_5	Biochemical oxygen demand mg O_2/l removed over 5 days	ML^{-3}
L_t	BOD remaining at time t, mg O_2/l	ML^{-3}
K_1	BOD reaction rate constant, per day	T^{-1}
L_0	Ultimate BOD, i.e. the result of a BOD test which is extended beyond 20 days, mg/l	ML^{-3}
k_1	BOD reaction rate constant to base 10, day^{-1} . $k_1 = 0.43K_1$	T^{-1}
L_N	BOD of nitrification stage, mg O_2/l	ML^{-3}
L_{NO}	Ultimate BOD for nitrification stage, mg O_2/l	ML^{-3}
K_N	Nitrification reaction rate constant, day^{-1} to base e	T^{-1}
k_N	Nitrification reaction rate constant, day^{-1} to base 10	T^{-1}
BOD_N	BOD exerted by nitrifying organisms, mg/l	ML^{-3}
t_N	Delay time in nitrification reaction when nitrifying organisms are at low and ineffective concentrations, day	T

MPN	Most probable number of a particular bacterial type,/100 ml	L^{-3}
r	Hydraulic mean depth (hydraulic radius) A/P, m	L
h	Loss of head in pipeline length, m/m	–
l	Pipeline length, m	L
f	Friction coefficient, dimensionless	–
g	Gravitational constant, m/s^2	LT^{-2}
C_z	Chezy coefficient, m$^{1/2}$/s	$L^{1/2}T^{-1}$

1.1 General

Wastewaters are often classified as stormwater (surface water and runoff from roofs, paved areas, etc.), sullage (water from households, excluding that from WCs) and foul water (water contaminated by faecal material). Separation of the stormwater from the other two types enables flexibility in storage treatment and disposal methods to be maximized. Sullage and foul sewage are rarely separated.

1.2 Stormwater sewerage

Storms vary in intensity (mm/h), duration (h or min) and frequency (number of storms of given intensity per 10 years, averaged over a long period).

The Bilham [1] formula, often used in the UK, relates these parameters thus:

$$T = 1.25t\,(0.0394w + 0.1)^{-3.55} \qquad (1.1)$$

where w is the total rainfall (mm) over duration t (h), T is the number of storms of this intensity (w/t) expected in 10 years. If $T = 10$, the storm is said to have a return period of one year or it is sometimes referred to as a '1 year storm'. Thus the return period of 1 year means that a storm of given duration t and intensity i will be followed by a storm of duration t and intensity $\geqslant i$, at an interval which will average out at 1 year. Nonetheless, since this is a statistical definition, there exists a finite probability that two or more such storms may occur at much closer intervals.

The Bilham formula can be used to construct the intensity–duration curve for storms with particular return periods as follows:

Example 1.1

For a storm likely to occur 10 times in 10 years ($T = 10$) and for a duration of 2 h ($t = 2$); then

Calculation

$$10 = 1.25 \times 2\,(0.0394\,w + 0.1)^{-3.55}$$

the overall rainfall

$$w = \frac{1}{0.0394}\left[\left(\frac{1.25 \times 2}{10}\right)^{1/3.55} - 0.1\right] = 14.64 \text{ mm}$$

thus the intensity (mm/h) = 14.64/2 = 7.32 mm/h □□□

Table 1.1 lists further values for w, w/t and t and these are plotted in Fig. 1.1.

Table 1.1 Duration, total rainfall and intensity for a storm with a 1 year return period ($T = 10$)

Duration (h)	Total rainfall (mm)	Intensity (mm/h)
0.1	4.85	48.5
0.5	9.08	18.2
1.0	11.59	11.6
1.5	13.30	8.9
2.0	14.64	7.3
3.0	16.72	5.6
4.0	18.34	4.6

Suggestion: plot log duration (h) versus log intensity (mm/h) for a storm with a return period of 1 year. Note that the curve can be *approximated* by a straight line plot with the relationship

$$\log_{10}t = -1.46 \log_{10}R + 1.56 \tag{1.2}$$

where R is the average intensity (mm/h) over duration t (h). The intensity versus duration curve for a 10 year return period storm ($T = 1$) has been added to Fig. 1.1. It is clear that a 10 year return period storm of a given duration produces approximately twice the rainfall intensity of a storm with a return period of 1 year and the same duration. In practice, a sewer which has been designed to cope with a 1 year return period storm may cope with a 1 in 10 year storm since flows will back up in manholes etc. and pressurize or surcharge the sewer, thus increasing the discharge rate.

1.2.1 Ministry of Health formulae

Apart from the Bilham formula, two other empirical formulae [2] are commonly used in the UK to determine rainfall intensity. For storms of duration 10 to 20 min

$$R = \frac{750}{t_m + 10} \text{ mm/h} \tag{1.3}$$

and for durations of 20 to 120 min

$$R = \frac{1000}{t_m + 20} \text{ mm/h} \tag{1.4}$$

where t_m is the duration in minutes. These formulae give similar intensities to the Bilham formula when $T = 10$, i.e. for a storm of 1 year return period.

Example 1.2
For a 1 year return period storm, calculate the intensities for storms of duration 6, 30 and 120 min using both the 'Ministry' and Bilham formulae.

Calculation

From equation (1.3) when $t_m = 6$ min; $R = \dfrac{750}{6 + 10} = 46.9$ mm/h

From equation (1.4) when $t_m = 30$ min; $R = \dfrac{1000}{30 + 20} = 20.0$ mm/h

when $t_m = 120$ min; $R = \dfrac{1000}{120 + 20} = 7.1$ mm/h

The Bilham formula (for $T = 10$ and t in h) gives intensities of 48.5, 18.2 and 7.3 mm/h for durations of 6, 30 and 120 min, respectively.

□□□

Although the Bilham and Ministry of Health formulae are based on UK practice the general form of the intensity–duration curves in Fig. 1.1 is found worldwide. A design problem occurs when a choice of storm frequency is made in order to provide storm sewers. Many factors influence this choice, including the accuracy of and the confidence in past rainfall records, the cost to life and property of an occasional flood and the cost of the sewer installation. However, assuming that a frequency can be chosen for design purposes, the maximum rate at which stormwater must be removed from the catchment area must be determined.

1.2.2 Time of concentration

If rain starts to fall over an area A_c and is removed in a sewer of length l (m), then after some initial delay, referred to as the entry time t_e (min), stormwater from the area closest to the outlet of the sewer will reach the sewer outlet. The discharge at the outlet will increase until eventually the whole area A_c is contributing. For the whole area to contribute to the flow, the storm must last until water from the areas furthest from the sewer outlet has entered the sewer

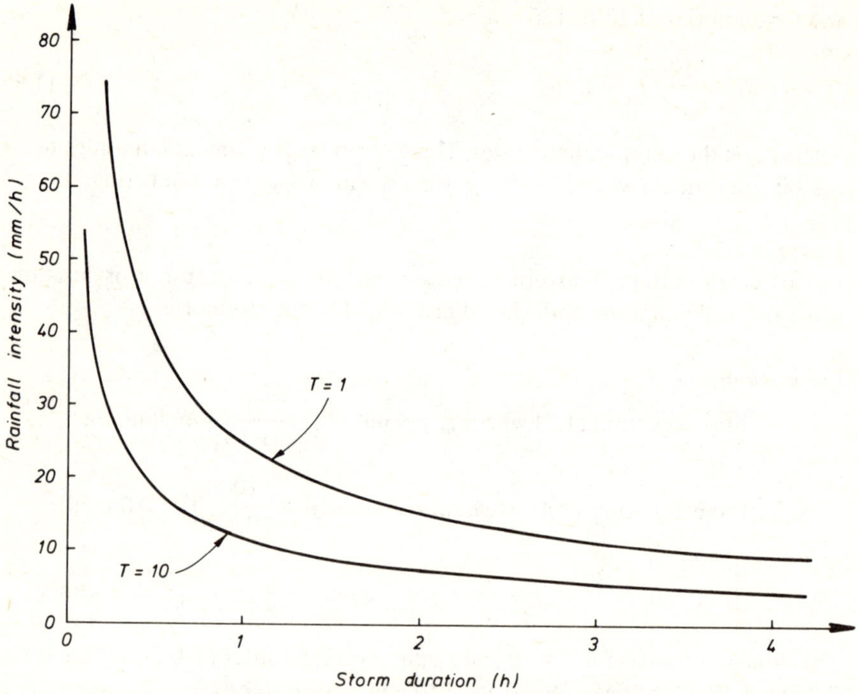

Fig. 1.1 Rainfall intensity versus duration curves

and flowed the length of the sewer. If the storm ceases before this time has elapsed the area closest to the sewer outlet would cease to contribute before the flow from the furthest point arrived at the outlet. The time to flow the length of the sewer (at full bore) plus the time of entry is the 'time of concentration t_c' for the area, i.e.

$$t_c = \frac{1}{60V_F} + t_e \tag{1.5}$$

where V_F is the full bore velocity (m/s) and t_c and t_e are in minutes. t_e is estimated in 0.5 min intervals[2].

The greatest flow rate at the sewer discharge point is achieved when the storm duration equals the time of concentration. It is interesting to consider the following three storms.

Example 1.3

For three storms X, Y and Z of durations $t_X = t_c/2$; $t_Y = t_c$ and $t_Z = 2t_c$ all of which fall on the 1 year return period storm curve in Fig. 1.1, show that storm Y produces the greatest discharge for a given catchment area. It should be noted that an assumption is made that the flow to the sewer Q (l/h) is equal to the impermeability of the area C, the intensity of rainfall i (mm/h) and the contributing area A_c(m^2), i.e.

$$Q = CiA_c \tag{1.6}$$

Calculation
Consider that for storm Y, at the end of the storm, water from all the catchment is reaching the sewer outlet. The discharge rate Q_Y is then $i_Y CA_c$. For storm Z, the maximum discharge Q_Z occurs when all the area is contributing, i.e. after t_c and up to 2 x t_c min. The discharge then is $Q_Z = i_Z CA_c$. But from Fig. 1.1 a storm with duration $t_Z > t_Y$ for the same return period must have intensity $i_Z < i_Y$. Therefore $i_Z CA_c < i_Y CA_c$, i.e. the maximum discharge rate for storm Z is less than that for storm Y. For storm X, only the contribution from approximately half the area would reach the discharge point by the time the storm ends. Thus half the area only is contributing and the discharge Q_X

$$Q_X = i_X C(A_c/2)$$

from inspection of Fig. 1.1, for a 1 year return period storm, halving the duration cannot produce twice the intensity; therefore $i_X/2 < i_Y$ and $Q_Y < Q_X$

□ □ □

An alternative approach to show that a storm W of duration $t_W = kt_c$ (where $k < 1$), produces less flow in the sewer than one where $t_Y = t_c$, is based on an approximation to Bilham's formula for a 1 year return period[3], i.e.

$$R = \frac{11.5}{t^{0.625}} \text{ mm/h} \tag{1.7}$$

The flow due to storm W is $Q_W = Ci_W kA_c$ and for storm Y is $Q_Y = Ci_Y A_c$. Now

$$i_W = \frac{11.5}{(kt_c)^{0.625}} = \frac{11.5\,(k^{-0.625})}{t_c^{\,0.625}}$$

and

$$i_Y = \frac{11.5}{t_c^{\,0.625}}$$

Thus

$$\frac{Q_W}{Q_Y} = C\frac{11.5}{t_c^{\,0.625}}\,k^{0.375}A_c \Big/ C\frac{11.5}{t_c^{\,0.625}}A_c$$

and for $k < 1$ the value of $Q_W < Q_Y$

1.2.3 The rational method

The Lloyd Davies[4] or rational method of determining the stormwater sewer flow from an area is based on two criteria:

(a) that maximum flow occurs when the storm duration is equal to t_c, and
(b) that $Q = CiA_c$ as before.

The value of i is more specific than R since it is the average intensity of rainfall over the time t_c.

The impermeability coefficient not only varies with type of surface and surface slope but also increases with duration of the storm. In some countries values of C are increased by 0.10 to 0.20 after the first 50 mm rainfall has fallen, whereas in other countries the value of C is varied according to the return period of the storm. Table 1.2 lists some illustrative values for C, however, local knowledge is invaluable and is usually followed.

Table 1.2 Impermeability coefficients $C = Q/iA_c$

Type of surface	C*
Lawns	0.10–0.20
Streets	0.70–0.95
Residential areas (flat)	0.30–0.50
(hilly)	0.50–0.70
Urban areas	0.70–0.90
Industrial areas (light)	0.55–0.65
(heavy)	0.60–0.80

* Values are for storms with a 5 year return period.

The rational method for sizing the sewer for a particular area is applicable for sewers <600 mm diameter. Above this diameter a computer based method such as that of the Transport and Road Research Laboratory[5] is preferable. The rational method involves an iterative (repeating) calculation. A formula for the velocity of flow in pipes must be assumed at this point; this is given in Section 1.3. Table 1.3 and the nomogram in Appendix 2 should be noted.

Example 1.4

Consider two areas served in series by a continuous sewer. Area 1 is 1.0 ha with an overall coefficient of impermeability C of 0.35 (residential) and has a ground slope of 1 in 400. Area 2 is again 1.0 ha with $C = 0.25$ with a slope of 1 in 300. Choose sewer diameters for the two areas given that the sewer length in area 1 is 150 m, for area 2 is 190 m and the storm return period is 1 year.

Table 1.3 Minimum gradients for pipes of given diameter*

Pipe diameter (mm)	Minimum gradient (m/m)	Flow rate (m^3/s)
100	0.0111	0.006
150	0.0064	0.013
200	0.0044	0.024
225	0.0038	0.030
250	0.0033	0.037
300	0.0026	0.053
375	0.0019	0.083
450	0.0015	0.120
525	0.0012	0.163
600	0.0010	0.212

* Based on a self cleansing velocity of 0.75 m/s and a Manning's coefficient $n = 0.012$ (see Section 1.3).

Calculation

Since the velocity in a sewer should be $\geqslant 0.75$ m/s an approximate value for t_c in area 1 is given by assuming $V_F = 0.85$ m/s. A value for t_e of 2.5 min has been chosen to illustrate the method.

From Equation (1.5)

$$t_c \text{ guessed} = \frac{150}{60 \times 0.85} + 2.5 = 5.4 \text{ min}$$

From Equation (1.3)

$$i = \frac{750}{10 + 5} = 50 \text{ mm/h}$$

The effect of rounding the time down to 5.0 min is to increase i. The flow in the sewer in area 1 is Q_1, i.e.

$$Q_1 = CiA = \frac{0.35 \times 50 \times 10\,000 \times 10^{-3}}{3600} \text{ m}^3/\text{s}$$

$$= 0.049 \text{ m}^3/\text{s}$$

From Table 1.3, a sewer of 300 mm diameter laid to a self cleansing gradient of 0.0026 (1 in 385) would be satisfactory and would carry 0.053 m^3/s. Other alternatives would be a 250 mm diameter sewer laid to a slope of 0.007 (1 in 143) (see Section 1.3 or nomogram A2.1 in Appendix 2). However this latter choice would increase V_F to 1.08 m/s; thus diminishing t_c to $150/(1.08 \times 60) + 2.5$ or 4.8 min and therefore increasing i by a small amount. Savings in cost of pipe would have to be offset against the increased excavation depth caused by the slope of 1 in 143.

Assuming the 300 mm diameter pipe is chosen, the value of V_F for the sewer when carrying 0.049 m^3/s is approximately (see Section 1.3) that of full bore flow, i.e. 0.75 m/s. The use of this value for V_F would increase t_c to 5.8 min and reduce i to 47 mm/h with consequent reduction in Q_1 to 0.046 m^3/s.

Area 2: For a sewer length of 190 m, estimating V_F as 0.85 m/s leads to a flow time in this length of 3.7 min. Adding the value of t_c for area 1 gives a total time t_c for the combined area of 3.7 + 5.8 = 9.5 min. Therefore

$$i = \frac{750}{19.5} = 38 \text{ mm/h}$$

The sum of $C \times A$ for the two areas is $0.35 \times 10^4 + 0.25 \times 10^4$ or 0.6×10^4 m^2. Therefore the combined flow $Q_{1+2} = (38/3600) \times 0.6 \times 10^4 \times 10^{-3}$ m^3/s, i.e.

$$Q_{1+2} = 0.063 \text{ m}^3/\text{s}$$

This could be carried by a 375 mm diameter pipe laid to a self cleansing slope of 0.0019 (1 in 526) but since the ground slopes at 0.0033 (1 in 300), a 300 mm diameter pipe laid to a slope of 0.0038 (1 in 260) would also suffice. At this slope the velocity would be 0.92 m/s which would reduce t_c overall to

$$t_c = \frac{190}{0.92 \times 60} + 5.8 \text{ min or } 9.2 \text{ min}$$

which would again result in i being 39 mm/h. This latter choice may therefore be cheaper than the 375 mm diameter pipe.

With some experience and intuitive choice of sewer sizes, the whole of the above example can be reduced to two or three rows of tabulated figures.

□□□

1.3 Flow in pipes

For unpressurized pipe flow conditions, Manning's formula is

$$V = \frac{r^{2/3} S^{1/2}}{n} \text{ m/s} \tag{1.8}$$

where V is the velocity of fluid (average) in the pipe (m/s); r is the hydraulic mean depth or hydraulic radius (m), i.e. the area of flow A(m^2) divided by the length of wetted perimeter P (m), S is the slope of the sewer (m/m), and n is a frictional coefficient which varies with the material of which the sewer is constructed and its condition. For most sewers a value of 0.012 is used.

A common way of expressing V is

$$V = \frac{1}{0.012}\left(\frac{A}{P}\right)^{2/3} S^{1/2} \tag{1.9}$$

where A is the cross-sectional area of flow (m^2).

1.3.1 Discharge and velocity versus depth of flow

From Fig. 1.2 for a pipe of diameter d:

$$\frac{d}{2}\cos\theta = \frac{d}{2} - Hd \tag{1.10}$$

where H is the fractional depth of flow, or

$$\theta = \cos^{-1}(1 - 2H) \tag{1.11}$$

The area of flow for a depth of flow $H \times d$ is, for θ in radians;

$$A = \frac{\theta d^2}{4} - \frac{d}{2}\sin\theta . \frac{d}{2}\cos\theta \tag{1.12}$$

since $(d/2)\sin\theta\,(d/2)\cos\theta$ is the shaded triangle in Fig. 1.2, thus

$$A = \frac{d^2}{4}(\theta - \sin\theta\cos\theta) \tag{1.13}$$

the wetted perimeter

$$P = \theta d \tag{1.14}$$

therefore

$$r = 0.25\,\frac{d}{\theta}(\theta - \sin\theta\, .\, \cos 0) \tag{1.15}$$

From the Manning formula, the velocity of flow V_H in a pipe with depth of flow H, is related to the velocity V_F in the same pipe at full flow as the 2/3 power of

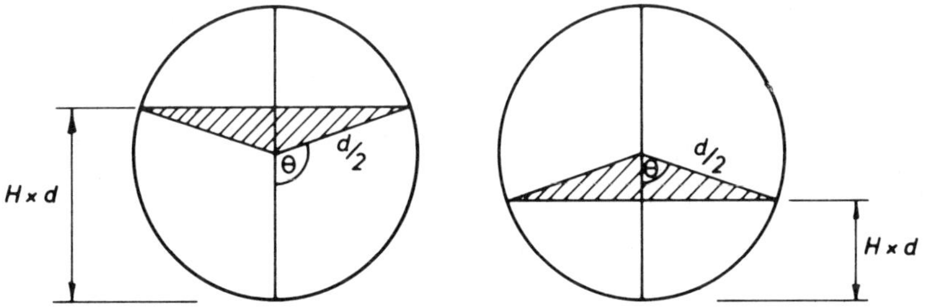

Fig. 1.2 Area of flow in pipes

the ratio of the two values of A/P, i.e.

$$\frac{V_H}{V_F} = \left[\frac{0.25(d/\theta)\,(\theta - \sin\theta\cos\theta)}{0.25d}\right]^{2/3} \tag{1.16}$$

or

$$\frac{V_H}{V_F} = \left[\frac{1}{\theta}(\theta - \sin\theta\cos\theta)\right]^{2/3} \tag{1.17}$$

Since the discharge or flow in the pipe Q (m^3/s) is equal to $V \times A$; the ratio of flow Q_H at depth of flow H to the flow Q_F when the pipe is full, is given by

$$\frac{Q_H}{Q_F} = \left[\frac{1}{\theta}(\theta - \sin\theta\cos\theta)\right]^{2/3} \times \left[\frac{(d^2/4)\,(\theta - \sin\theta\cos\theta)}{\pi(d^2/4)}\right] \tag{1.18}$$

or

$$\frac{Q_H}{Q_F} = \frac{1}{\pi}\left(\frac{1}{\theta}\right)^{2/3} (\theta - \sin\theta\cos\theta)^{5/3} \tag{1.19}$$

Example 1.5

For a depth of flow of $0.2d$, i.e. $H = 0.2$, calculate the velocity and flow as a fraction of the full bore values.

Calculation

$$\theta = \cos^{-1}(1 - 2 \times 0.2) = 0.927 \text{ radians}$$

$$\frac{V_{0.2}}{V_F} = \left[\frac{1}{0.927}(0.927 - 0.480)\right]^{2/3}$$

i.e.

$$\frac{V_{0.2}}{V_F} = 0.62$$

Also

$$\frac{Q_{0.2}}{Q_F} = \frac{1}{\pi}\left(\frac{1}{0.927}\right)^{2/3} (0.927 - 0.799 \times 0.600)^{5/3}$$

thus

$$\frac{Q_{0.2}}{Q_F} = 0.088$$

□ □ □

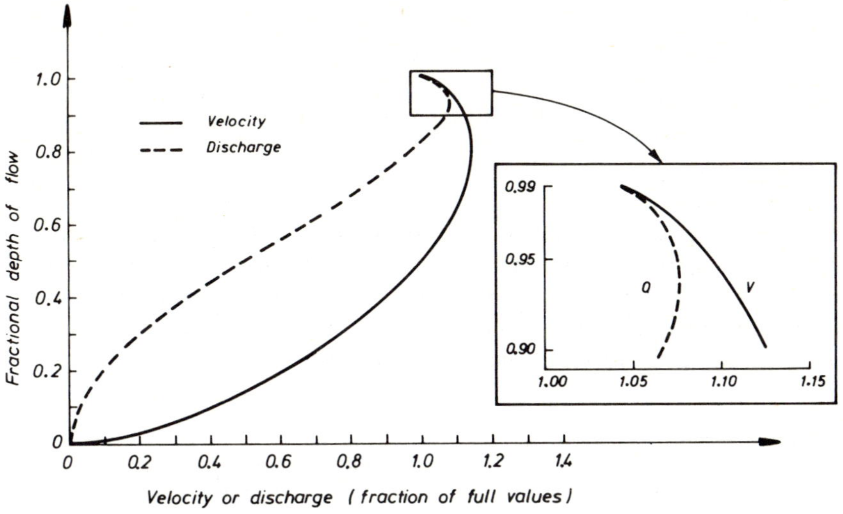

Fig. 1.3 Velocity and discharge versus depth of flow in pipe

Fig. 1.3 shows the variation of the ratios of Q_H/Q_F and V_H/V_F with depth of flow H. Two features are noteworthy, namely, the value of the velocity is the same for $H = 0.5$ as for full bore, and the pipe flow appears to be a maximum at $H = 0.94$.

When $H = 0.5$

$$\theta = \cos^{-1}(1 - 1) \text{ or } 1.571 \text{ radian}$$

$$\frac{V_{0.5}}{V_F} = \left[\frac{1}{1.571}(1.571 - 1.00 \times 0.00)\right]^{2/3}$$

i.e.

$$\frac{V_{0.5}}{V_F} = 1$$

This result could have been seen by inspection since the term A/P is precisely the same value for $H = 1$ and $H = 0.5$. Similarly, the ratio of the areas is seen to be

$$\frac{A_{0.5}}{A_F} = \frac{1}{2} \text{ so } \frac{Q_{0.5}}{Q_F} = 0.5$$

where A_F is the total cross-sectional area of the pipe. Thus the 1/2 full bore discharge is 50 % of that running full.

When $H = 0.94$

$$\theta = \cos^{-1}(1 - 1.88) = 2.65 \text{ radians}$$

$$\frac{Q_{0.94}}{Q_F} = \frac{1}{\pi}\left(\frac{1}{\theta}\right)^{2/3} [\theta - \sin\theta \cos\theta]^{5/3}$$

$$= 1.08$$

i.e. the discharge is at its maximum value of 108% of the full bore discharge.

This result appears to offer some advantage to the designer if the pipes were operated at a depth of flow of 0.94, however, instabilities in the flow tend to occur which reduce the pipe flow below the theoretical value. For this reason, pipes are assumed to carry maximum flow when running full.

1.3.2 Discharge and velocity versus slope for pipes running full

If n in the Manning formula is taken as 0.012, then for a pipe running full $r = d/4$ and for d in mm

$$V = 0.33\, d^{2/3}\, S^{1/2} \text{ m/s} \tag{1.20}$$

Also, since

$$Q = VA \text{ m}^3/\text{s} \tag{1.21}$$

then

$$Q_F = 0.33\, d^{2/3}\, S^{1/2} \frac{\pi d^2}{4} \times 10^{-6} \text{ m}^3/\text{s} \tag{1.22}$$

and

$$Q_F = 2.6 \times 10^{-7}\, d^{8/3}\, S^{1/2} \tag{1.23}$$

Example 1.6

For a 300 mm diameter pipe laid to a slope of 1 in 400(0.0025) calculate the flow rate and velocity when running full

Calculation

$$Q_F = 2.6 \times 10^{-7} \times 300^{8/3}(0.0025)^{1/2}$$

i.e.

$$Q_F = 0.052 \text{ m}^3/\text{s}$$

and

$$V_F = 0.74 \text{ m/s}$$

Equations (1.20) and (1.23) are usually referred to as the Crimp and Bruges formulae. It is clear from Table 1.3 that the minimum slopes quoted for particular pipe diameters are calculated to give a full bore velocity of 0.75 m/s when $n = 0.12$.

□ □ □

1.3.3 Other formulae for flow in pipes

One of the commonest formula for flow in pipes which are pressurized is the D'Arcy–Weisbach formula which is usually given in the form below:

$$h = \frac{flV^2}{2gd} \tag{1.24}$$

where h is the head loss (m) in length of pipe l (m); V is the velocity of flow in the *full* pipe (m/s); g is the gravitational constant (m/s^2); d is the pipe diameter (m) and f is a dimensionless friction coefficient.

For a partially full pipe, a more usual version is

$$h = f\,\frac{V^2 l}{8gr} \tag{1.25}$$

where r is the hydraulic mean depth as before. When the pipe is flowing full, $r = d/4$ and Equation (1.25) becomes Equation (1.24). The value of f may be determined from a knowledge of the equivalent roughness of the pipe and the Reynolds number by means of the Moody diagram (Appendix 2). However, if the Manning coefficient n is known, an equivalent value of f can be found as follows. Squaring both sides of the Manning formulae gives

$$V^2 = \frac{r^{4/3}S}{n^2} \tag{1.26}$$

or

$$S = \frac{V^2}{r}\,\frac{n^2}{r^{1/3}} \tag{1.27}$$

For unsurcharged or unpressurized uniform flow

$$S = \frac{h}{l} \tag{1.28}$$

Now, from the D'Arcy–Weisbach equation

$$\frac{h}{l} = \frac{V^2}{r}\,\frac{f}{8g} \tag{1.29}$$

Thus

$$f = \frac{8gn^2}{r^{1/3}} \tag{1.30}$$

Example 1.7

A 150 mm diameter sewer has a Manning's *n* coefficient of 0.012 and is laid to a slope of 1 in 120. Use the D'Arcy–Weisbach equation to calculate the volumetric flow when running half full.

Calculation

From the D'Arcy–Weisbach equation for a pipe flowing partially full

$$V = \left[\frac{h}{l}\left(\frac{8gr}{f}\right)\right]^{1/2} \tag{1.31}$$

or

$$V = \left[\frac{8grS}{f}\right]^{1/2} \tag{1.32}$$

also

$$f = \frac{8gn^2}{r^{1/3}}$$

thus

$$V = \left(\frac{8gSr^{4/3}}{8gn^2}\right)^{1/2} \quad \text{or} \quad \frac{r^{2/3}}{n}\sqrt{S} \tag{1.33}$$

Since $Q = VA$ and for $r = (\pi d^2)/(4\pi d)$ or $d/4$ at 1/2 depth of flow, then

$$Q = \frac{\pi d^2}{8}\,\frac{r^{2/3}}{n}\sqrt{S} \tag{1.34}$$

or

$$Q = \frac{\pi d^2}{n8}\left(\frac{d}{4}\right)^{2/3}\sqrt{S} \tag{1.35}$$

It should be noted that this is, of course, identical to the Manning formula for *d* in m; namely

$$A \times V = \frac{\pi d^2}{8} \times \frac{1}{n}\left(\frac{d}{4}\right)^{2/3}\sqrt{S}$$

The flow therefore

$$= \frac{\pi\,(0.15)^2}{8 \times 0.012}\left(\frac{0.15}{4}\right)^{2/3} \sqrt{(8.3 \times 10^{-3})}$$

$$= 7.5 \times 10^{-3}\ \text{m}^3/\text{s} \ \text{ or } \ 7.5\ \text{l/s}$$

□ □ □

Another equation which is often used in pipe flow calculations is the Chezy formula, which can be written as

$$V = C_z\sqrt{(rS)} \tag{1.36}$$

for an unsurcharged pipe where C_z is a coefficient. It can be readily shown that

$$C_z = \left(\frac{8g}{f}\right)^{1/2} = \frac{r^{1/6}}{n} \tag{1.37}$$

A form of the Moody diagram and instructions for its use in the determination of f, together with a pipe flow nomogram, is included in Appendix 2.

1.4 Foul sewerage

Foul sewerage calculations are usually based on the concept of dry weather flow (DWF) which is the average daily flow in the sewer after several days during which the rainfall has not exceeded 0.25 mm. A formula is often used where direct measurement cannot be made, i.e.

$$\text{DWF} = N \times G + I + T \quad \text{l/day} \tag{1.38}$$

where N is the population (number) served by the sewer; G is the average daily per capita water usage (l/cap . day); I is the average (dry weather) infiltration into the sewer owing to poor joints or pervious materials etc. (l/day) and T is the average trade waste discharge in l/day.

N is usually determined from census data with allowance made for future growth, G is approximately equal to the per capita daily water supply, I can be estimated as 1/h per km sewer for various ages of sewer and T may be estimated from local figures for industrial water supplies which are metered.

1.4.1 Dry weather flow variations

The instantaneous inflow to a sewage works is generally expressed as a multiple of DWF, and the preliminary works are usually designed to divert or give different treatment to particular flow fractions. The multiple of DWF which has to be treated decreases with increasing size of the sewerage system. In small units, this multiple may be extremely high.

Example 1.8

A septic tank serves a house containing 4 persons discharging 180 1/cap . day by means of a 100 mm (4 in) diameter sewer pipe. If surface water is totally excluded the following multiples of DWF occur:

$$\text{DWF} = \frac{4 \times 180}{8.64 \times 10^4} \text{ l/s} = 0.0083 \text{ l/s}$$

At night the flow will be almost zero. When 1 x 10 l WC is flushed in 10 s the flow is 1 l/s or 120 DWF. When 1 x 50 l bath is discharged over 30 s the flow is 201 DWF.

□ □ □

Where separate foul sewers are constructed the design flow on which the sewer is based is often 4 DWF or 6 DWF. If combined foul and surface water sewers are constructed the storm water can be shown to be the major design criterion (in the following calculations trade wastes and groundwater infiltration into the sewer are assumed to be negligible).

Example 1.9

In a mainly residential area of 100 persons/ha the per capita daily water usage is 220 l/cap . day. Compare the foul sewage flow with a stormwater flow produced by an intensity of rainfall of 40 mm/h when $C = 0.35$ for the district.

Calculation

For the foul sewage

$$\text{the DWF} = 220 \times 100 \text{ l/ha . day or } 22.0 \text{ m}^3\text{/ha . day}$$

At a peak flow of 6 DWF this is 1.5 l/s per ha. For the storm flow

$$Q = 0.35 \times \frac{40}{3600} \times 10^4 \times 10^{-3} \text{ m}^3\text{/s}$$

i.e.

$$Q = 0.039 \text{ m}^3\text{/s or 39 l/s}$$

The ratio is therefore 1:26 of foul sewage to storm water.

□ □ □

When calculating the diameter of a foul water sewer on a totally separate system, an estimate must be made of the inflow to the sewer of infiltration water owing to poor joints or pervious materials. A figure of 120 l/h per km of *new* sewer has been used in the UK with somewhat larger 200 l/h km values being common in the USA[6]. Foul sewers are considerably smaller than stormwater sewers.

Example 1.10

Three sewers each 300 m long are to be constructed to each serve 75 houses. The occupancy rate is 3.7 per house and the water consumption is 240 l/cap . day. The three sewers feed a larger sewer 200 m long which has additionally to serve 50 houses. Determine suitable pipe diameters and slopes when the ground slope is 1 in 100.

Calculation

For each of the 3 sewers the average domestic flow is

$$\text{DWF (less infiltration)} = 240 \times 3.7 \times 75 \text{ l/day}$$

Ignoring the infiltration water, 6 DWF is 4.6 l/s. From Table 1.3, a 100 mm sewer laid at 0.011 slope (1 in 91) would accommodate 5.9 l/s.

At 6 DWF the ratio of Q_H/Q_F would be 0.78. From Fig. 1.3 H would be 0.67 and the corresponding velocity would be 1.1 x the full bore velocity of 0.75 m/s.

Bartlett[7] has suggested that the sewer should be designed to reach a self cleansing velocity once a day and he has also suggested that 2 DWF is the maximum flow that can be relied upon to occur at least once a day. Therefore the sewer should produce a flow velocity of $\geqslant 0.75$ m/s at 2 DWF.

Now 2 DWF = 1.5 l/s. Therefore, using Fig. 1.3

$$\frac{Q_H}{Q_F} = 0.25; \; H = 0.37 \text{ and } \frac{V_{0.37}}{V_F} = 0.89$$

Therefore V_H at 2 DWF = 0.89 x 0.75 or 0.67 m/s, which is unsatisfactory. Increasing the slope to 0.020 (1 in 50) increases Q_F to

$$Q_F = 5.9 \times \frac{\sqrt{0.020}}{\sqrt{0.011}} \text{ or } 8.0 \text{ l/s}$$

Thus at 2 DWF (Equation (1.20))

$$\frac{Q_H}{Q_F} = 0.19; H \text{ is } 0.29 \text{ and } V_{0.29}/V_F \text{ is } 0.77$$

Thus at 2 DWF

$$V = 0.77 \times 0.33 \times 100^{2/3} \sqrt{0.02}$$

$$= 0.77 \text{ m/s}$$

This indicates quite clearly the considerable problems, with even the minimum size of sewer, in achieving self cleansing velocities.

For a 150 mm diameter sewer at 2 DWF

(*a*) At slope 0.0064; $Q_H/Q_F = 0.11$; $H = 0.2$; $V_{0.2}/V_F = 0.6$ and $V_{0.2} =$ 0.45 m/s.
(*b*) At slope 0.02; $Q_H/Q_F = 0.06$; $H = 0.15$; $V_{0.15}/V_F = 0.5$ and $V_{0.15} =$ 0.66 m/s.

Because of this problem of achieving self cleansing velocities, it has been found necessary to connect some of the surface water drains on the houses at the head of the sewer to the foul sewer. This at least ensures cleansing in times of storm, and enables 150 mm diameter sewers to be used; the use of 100 mm diameter sewers being unpopular owing to potential blockages.

For the sewer after the junction of the three minor sewers

$$6 \text{ DWF} = 4.6 \times 3 + 3.0 \text{ l/s or } 16.8 \text{ l/s}$$

A 150 mm diameter sewer at a slope 0.012 (1 in 83) would carry 18.1 l/s at a velocity of 1.02 m/s and at 2 DWF. $Q_H/Q_F = 0.31$; $H = 0.37$ and $V_{0.37}/V_F =$ 0.86; thus $V_{0.37} = 0.88$ m/s. A 150 mm diameter sewer at a slope of 0.012 (1 in 83) is therefore satisfactory.

□ □ □

1.5 Sewage characteristics

The pollutional loading of wastewaters from individual trades, farms etc. may be expressed in terms of kg/day of particular compounds. However, for mainly domestic wastes, the three most common parameters used to describe the waste are the biochemical oxygen demand (BOD), the suspended solids (SS) and the bacteriological quality.

1.5.1 Biochemical oxygen demand (BOD)

This is the oxygen used up by the bacterial oxidation of organic pollutants (carbonaceous BOD) or ammonia (nitrogenous BOD) usually determined under standard conditions of incubation at 20 °C over 5 days. In this case the nitrogenous BOD will be small. A BOD of 10 mg/l means that 1 l of waste will require 10 mg of O_2 when incubated at 20 °C for 5 days. The sample is incubated in a sealed bottle (250 to 300 ml); the initial dissolved oxygen concentration (measured on a duplicate sample) less the final dissolved oxygen concentration (measured on the incubated sample) gives a measure of the BOD. The necessary bacteria and nutrients may be present naturally or may be added[8].

Example 1.11

A BOD test was carried out on three samples of river water. All gave a dissolved oxygen initial reading of 7 mg/l. The final readings after 5 days incubation at 20 °C were 0, 3 and 7 mg/l. Calculate the three BOD values.

Calculation
For the sample having a final dissolved oxygen content of 3 mg/l, the BOD is $7 - 3 = 4$ mg/l.

For the sample with a final DO value of 0 mg/l, no result can be stated except that the BOD is $\geqslant 7$ mg/l.

In the case of the third sample the apparent result is a BOD of 0 mg/l. However, since this result would also be given for a sample that contained bactericidal substances, unless it was known that the sample source was unpolluted, the sample should be examined further.

□ □ □

In practice if the BOD_5 is greater than $\cong 7$ mg/l dilutions of the sample must be used, e.g. if the second result was obtained from a sample diluted 1/100, the BOD_5 would be $(7 - 3)/0.01$. A detailed description of the analytical method is given elsewhere [8].

The kinetics of BOD oxidation depend upon the particular organics present, however, for domestic wastes a first-order reaction rate is assumed:

$$\frac{dL}{dt} = -K_1 L \tag{1.39}$$

L is the BOD remaining or unoxidized (mg/l), t is time (d) and K_1 is a constant with units of T^{-1} (day^{-1}).
Integration gives

$$\frac{L_t}{L_0} = e^{-K_1 t} \tag{1.40}$$

or, more usually

$$\frac{L_t}{L_0} = 10^{-k_1 t} \tag{1.41}$$

L_0 is the initial BOD when $t = 0$. This is referred to as the ultimate BOD (mg/l), and approximates to the result of a BOD test carried out at 20 °C over >25 days.

The BOD_5 is the BOD *removed* over 5 days. Now after 5 days the BOD remaining L_5 is given by

$$L_5 = L_0 10^{-k_1 5} \tag{1.42}$$

But

$$L_0 = L_5 + BOD_5 \tag{1.43}$$

Therefore

$$BOD_5 = L_0(1 - 10^{-k_1 5}) \tag{1.44}$$

or, generally

$$\mathrm{BOD}_t = L_0(1 - 10^{-k_1 t}) \tag{1.45}$$

Example 1.12

A BOD test is carried out on domestic sewage which has a k_1 value at 20 °C of 0.16 day^{-1}. Calculate the BOD_5 values (the BOD exerted) as a fraction of the ultimate BOD.

Calculation

From Equation (1.45)

$$\frac{\mathrm{BOD}_5}{L_0} = 1 - 10^{-(0.16 \times 5)}$$

i.e.

$$\frac{\mathrm{BOD}_5}{L_0} = 0.84$$

i.e. the 5 day BOD is 84% of the ultimate BOD.

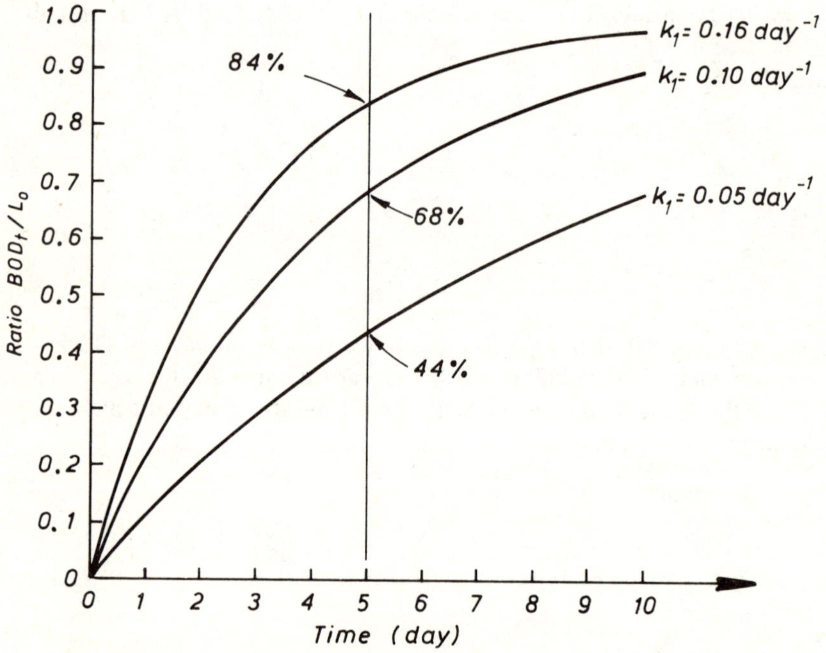

Fig. 1.4 BOD_t/L_0 for different values of k_1

Fig. 1.4 shows the ratios of BOD_t/L_0 for two other reaction rates, and indicates that at 20 °C $k_1 = 0.16$ day^{-1} the BOD_{10} is 97% of L_0.

□ □ □

(a) Effect of temperature
Within the viability range of the bacteria in the sample, the value of k_1 varies with temperature T °C:

$$k_T = k_{20}\ 1.047^{(T-20)} \qquad (1.46)$$

where k_T and k_{20} represent the values of k_1 at T °C and 20 °C, respectively.

Example 1.13
Assuming k_1 at 20 °C is 0.16 day^{-1}, calculate the ratios of BOD_t/L_0 for $T = 10°$ and $T = -5$ °C.

Calculation
For $T = 10$ °C

$$k_{10} = 0.16 \times 1.047^{(-10)}$$

$$= 0.10\ \text{d}^{-1}$$

and

$$\frac{BOD_t}{L_0} = 1 - 10^{(-0.10 \times 5)} = 0.68$$

or the BOD_5 is 68% of L_0.

□ □ □

Theoretically, even at –5 °C the value of k_1 is 0.05 day^{-1} and $BOD_t/L_0 = 0.44$. However, bacterial activity tends to diminish faster than the expression in Equation (1.46) would indicate even if freezing did not occur (see Section 9.2). It is inadvisable to use Equation (1.46) below 5 °C or above 30 °C without correcting for variations in L_0 with temperature as in Chapter 9, Equation (9.1).

(b) BOD as a conservative element
Since k_1 is small, the BOD concentration does not change appreciably during physical processes lasting several minutes. It can therefore be treated mathematically as a conservative element.

Example 1.14
A sewer carrying 0.3 m^3/s foul sewage with a BOD of 250 mg/l joins two other sewers carrying flows of 0.1 m^3/s and 0.2 m^3/s with BOD_5 of 379 mg/l and 157 mg/l, respectively. Calculate the BOD of the mixture.

Calculation

Flows are in the ratio 3:1:2.

Thus $(3 \times 250) + (1 \times 379) + (2 \times 157) = (3 + 1 + 2) \times BOD_{mixture}$

Therefore

$$BOD_{mixture} = 241 \text{ mg/l}$$

Also, since BOD is a measure of mg/l of O_2, the product of flow x concentration gives a mass mg/s which can be summed, e.g. for the above example

$$3 \times 10^2 \times 250 \text{ mg/s} + 10^2 \times 379 \text{ mg/s} + 2 \times 10^2 \times 157 \text{ mg/s}$$

$$= 0.14 \times 10^6 \text{ mg/s}$$

or

$$0.14 \text{ kg } BOD_5/\text{s}$$

(*c*) *Nitrification*

The term BOD in general calculations on loading, sewage treatment, effluent standards etc. usually means the 5 day 20 °C BOD exerted by the carbonaceous material. In most circumstances this is acceptable because the nitrosomonas and nitrobacter species of bacteria which are responsible for oxidizing $NH_3 \rightarrow NO_2^-$ and $NO_2^- \rightarrow NO_3^-$, respectively, are usually in very low concentrations in most samples of rivers, sewage, effluents etc. However, if the sample is taken after a lightly loaded percolating filter, the nitrogenous oxygen demand may be partially exerted within the 5 day period.

The nitrogenous kinetics are usually assumed to be first-order, i.e.

$$\frac{dL_N}{dt} = -K_N L_N \qquad (1.47)$$

where L_N is the nitrification BOD remaining (mg O_2/l) at time t(day); L_{NO} is the ultimate nitrification BOD (mg O_2/l); and K_N is a reaction constant (day^{-1}). Thus

$$BOD_N = L_{NO}(1 - 10^{-k_N t}) \qquad (1.48)$$

where BOD_N is the BOD due to the nitrification step and k_N is the value of K_N to base 10.

For a combined carbonaceous BOD_C plus nitrogenous BOD_N, the total BOD is

$$BOD = L_0(1 - 10^{-k_1 t}) + L_{NO}(1 - 10^{-k_N(t - t_N)}) \qquad (1.49)$$

where t_N is a delay time (day) before nitrifying organisms are effective.

Example 1.15
A BOD test is carried out for 15 days at 20 °C. The reaction rate constants k_1 and k_N are 0.16 day^{-1} and 0.1 day^{-1}, respectively, and L_0 and L_{NO} are 450 mg/l and 300 mg/l, respectively. The nitrifying organisms are assumed to be effective only from the time $t = 8$ days. Calculate the BOD exerted at days 5, 8, 12 and 15.

Calculation
Since BOD_N is ineffective until $t = 8$ days, then BOD_5 and BOD_8 are simply the carbonaceous demands.

$$\text{Total BOD}_5 = L_0(1 - 10^{-k_1 5}) = 450(1 - 10^{-0.16 \times 5}) = 379 \text{ mg/l}$$

$$\text{Total BOD}_8 = 450(1 - 10^{-0.16 \times 8}) = 426 \text{ mg/l}$$

For day 12

$$\text{Total BOD}_{12} = L_0(1 - 10^{-k_1 \times 12}) + L_{NO}(1 - 10^{-k_N(12-8)})$$

$$= 445 + 181$$

$$= 626 \text{ mg/l}$$

and total $BOD_{15} = 448 + 240$

$$BOD_{15} = 688 \text{ mg } O_2/\text{l}$$

□□□

Further discussion of nitrification is given in Chapters 4 and 5.

1.5.2 Suspended solids

The material in suspension in sewage commonly consists of inorganic grits, silts or clays, organic compounds such as bacteria, fats and greases, and a wide range of wastes from food cleaning and preparation processes. The total suspended material which will normally be removed by passage through a fine glass-fibre filter paper (GF/c) is classed as suspended solids SS.

Both BOD and SS concentrations in sewages are influenced by the daily per capita water usage. Present practice is therefore to assume a per capita production of BOD and SS in terms of kg/cap . day and to calculate expected concentrations in terms of local water usage.

Example 1.16
A small sewage works serves a population of 500 people whose per capita BOD_5 and SS production is 0.055 kg/day and 0.08 kg/day, respectively. Water consumption is metered at 200 l/cap . day. A small industry discharges an effluent of 5 l/s containing 100 mg/l BOD_5 and 200 mg/l SS over a period of 3 hours per day. Calculate the mean BOD_5 and SS concentrations in the works inlet.

Calculation

Domestic BOD = 0.055 x 500 kg = 27.5 kg

Industrial BOD = $5 \times 3600 \times 3 \times 100 \times 10^{-6}$ = 5.4 kg

Total = 32.9 kg BOD_5

Similarly the total SS = 50.8 kg

$$\text{The total flow} = (500 \times 200 + 5 \times 3 \times 3600) \times 10^{-3}\ m^3/day$$
$$= 154\ m^3/day$$

The mean concentrations are BOD 213 mg/l; SS 330 mg/l

□□□

1.5.3 Bacteriological standards

Some countries use bacteria as an indication of the pollutional state of the water sample. Certain bacteria derived from the gut of warm blooded animals are used as an indication of pollution, e.g. *faecal coliforms* (*f. coli.*). Assuming that a suitably specific test is available for the determination of a particular group of bacteria, the assessment of the number of organisms present is usually expressed as a most probable number, e.g. as MPN/100 ml sample. An example is the fermentation tube test for *faecal coliforms*[8]. Five tubes containing 10 ml of sample, five containing 1 ml and diluent, and five containing 0.1 ml and diluent, together with the necessary growth medium, are incubated for the specified time and the number of tubes showing positive reactions are determined. Further dilutions are necessary if all tubes show a positive result, since this is an open-ended result.

MPN values are usually read from tables, however, Thomas' formula[8] may be used:

$$\text{MPN/100 ml} = 100 \times \frac{\text{number of positive tubes}}{\left[\begin{array}{l}\text{volume sample in -ve tubes (ml)}\\ \times \text{ volume sample in all tubes (ml)}\end{array}\right]^{1/2}} \qquad (1.50)$$

Example 1.17

(1) An MPN assay for *f. coli.* is carried out on a stream using 5 x 10 ml, 5 x 1 ml and 5 x 0.1 ml samples. Three of the 10 ml, two of the 1 ml and one of the 0.1 ml samples showed positive results.

(2) A sewage, diluted 10 000 times was similarly assayed and showed positive results in four of the 10 ml, four of the 1 ml and none of the 0.1 ml sample tubes. Calculate the MPN for the river water and the sewage.

Calculation

For the river the positive results are 3 + 2 + 1 = 6. The volume of sample in negative tubes is $(2 \times 10 + 3 \times 1 + 4 \times 0.1)$ml = 23.4 ml. The total sample volume is (5 x 10 + 5 x 1 + 5 x 0.1)ml = 55.5 ml. Thus

$$\text{MPN/100 ml} = \frac{100 \times 6}{\sqrt{(23.4 \times 55.5)}}$$

$$= 17/100 \text{ ml}$$

For the sewage, the MPN is $\frac{100 \times 6}{(11.5 \times 55.5)}$

$$= 32/100 \text{ ml}$$

However the dilution factor is 10^4: the final MPN/100 ml for the sewage is 32×10^4/100 ml.

□ □ □

Note that this is a statistical technique and therefore the results are perhaps best expressed in terms of the spread of MPN between the lower and upper 95% confidence limits, e.g. for the river the spread is 5 to 46 per 100 ml and for the sewage 12 to 93 x 10^4/100 ml. Confidence limits for other results are given in the literature[8].

Questions

Question 1.1

Using the Bilham formula, find the maximum rate of stormwater runoff from an area of 2 ha with an impermeability of 0.5 for a storm with a 1 year return period. The time of concentration of the area is 1 h.

Question 1.2

A 300 mm diameter pipe supplies water from a reservoir to a town network at point A 2820 m distant. The TWL in the reservoir is at 184 m above a datum and the point A is at 150 m above the same datum. If the value for the D'Arcy–Weisback constant f for the pipe is 0.026, calculate the maximum pressure head at A when the pipe is delivering 53 l/s running full. Assume no drawdown of the reservoir level.

References

1 Bilham, E. G. (1935), Classification of heavy falls of rain in short periods. *British Rainfall.*
2 British Standard Code of Practice 2005 (1968), *Sewerage*. British Standards Institution, London.
3* Escritt, L. B. (1965), *Sewerage and Sewage Disposal*, 3rd Edn, C. R. Books Ltd., London.

4 Lloyd Davies, D. E. (1905), The elimination of storm water from sewerage systems. *Proc. Inst. Civil Engrs.*, **164**, 41–67.

5 Transport and Road Research Laboratory, Note No. 35 (1963), A guide for engineers to the design of storm water sewer systems. (See also Reference 2).

6* ASCE (joint publication) (1977), *Manual of engineering practice No. 36: Sewage treatment plant design*, New York.

7* Bartlett, R. E. (1976), *Surface water sewerage*, Applied Science Publications, London.

8 *Standard methods for the Examination of Water and Wastewater* (1976), 14th Edn, Joint publication, American Public Health Association, American Water Works Association and Water Pollution Control Federation, APHA, Washington, DC.

* Suggested further reading.

2
Preliminary treatment

Nomenclature

Symbol	*Definition and units*	*Dimensions*
DWF	Daily dry weather flow, m^3/day or 1/day	L^3T^{-1}
N	Population number	–
G	Per capita water consumption, 1/day or m^3/day	L^3T^{-1}
E	Industrial wastewater flow to sewer, 1/day or m^3/day	L^3T^{-1}
I	Infiltration flow into sewer (dry weather), 1/day or m^3/day	L^3T^{-1}
Q	General flow rate, m^3/s, 1/s or as stated	L^3T^{-1}
C	Empirical constant, $m^{1/2}$/s	$L^{1/2}T^{-1}$
h	Depth of flow over weir or through flume, m	L
V	Velocity in direction of flow, m/s	LT^{-1}
g	Gravitational constant, 9.81 m/s^2	LT^{-2}
B	Channel width, m	L
b	Width of flume throat or weir width, m	L
P	Height of weir crest above channel invert, m	L
h_s	Depth of flow over side weir, m	L
Q_{max}	Maximum flow to sewage works (old 6 DWF value)	L^3T^{-1}
W_s	Side weir width, m	L
m	Particle mass, kg	M
ρ_w	Fluid density, kg/m^3	ML^{-3}
ρ_p	Particle density, kg/m^3	ML^{-3}
A	Area or projected area of particle in horizontal plane, m^2	L^2
u	Particle velocity in vertical direction, m/s	LT^{-1}
C_D	Drag coefficient, dimensionless	–
R_e	Reynolds number, dimensionless	–
R_D	Relative density ρ_p/ρ_w or specific gravity	–
d	Particle diameter, m	L
ν	Kinematic viscosity of fluid, m^2/s	L^2T^{-1}
L	Length (e.g. of channel), m	L

H	Height or depth of flow, m	L
r	Hydraulic radius, m	L
S	Channel slope (fraction)	–
n	Coefficient varying with wall roughness, $s/m^{1/3}$	$TL^{-1/3}$
W	Channel width at water surface, m	L
K	Constant, dimensionless	–
f	Friction factor, dimensionless	–

2.1 Introduction

The inflow to a sewage works is dependent on the sewerage system. In a totally separate system where the minimum amounts of surface water are taken to the sewage works, the total inflow may receive the type of treatment shown diagrammatically in Fig. 2.1. In some parts of the world (e.g. the UK) it has been common to give up to three times dry weather flow, preliminary, primary and secondary treatment while flows above 3 DWF are treated as shown in Fig. 2.2.

The method used to define the value of 3 DWF and 6 DWF has varied, but a modern definition is [1]

$$3\ \text{DWF} = 3\,NG + I + 3E \tag{2.1}$$

$$6\ \text{DWF} = NG + I + 1360N + 3E \tag{2.2}$$

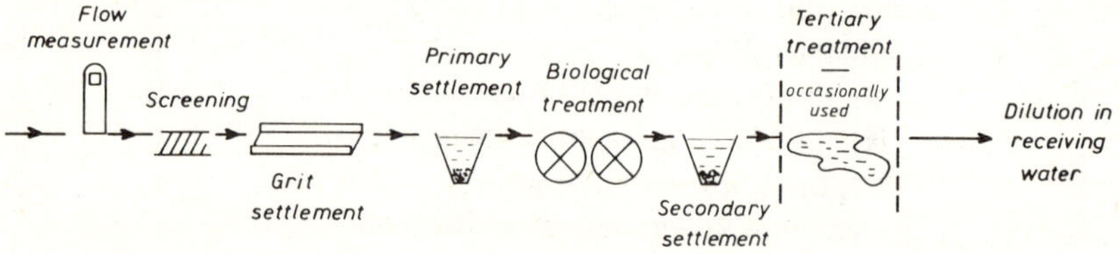

Fig. 2.1 Schematic sewage treatment diagram (flow receiving full treatment)

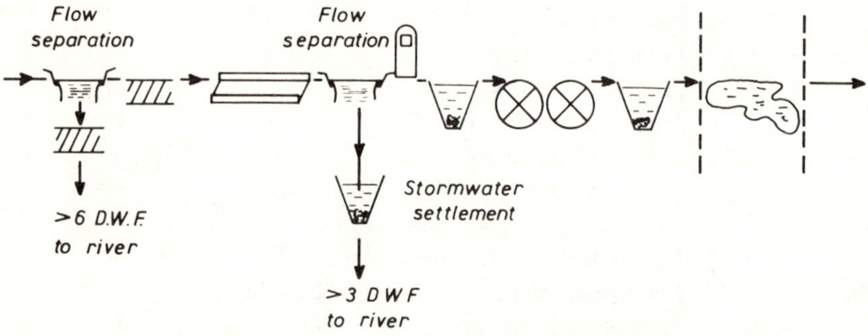

Fig. 2.2 Schematic sewage treatment flow diagram showing stormwater separation

where N is the population number, G is the daily per capita water usage (l/cap . day), I is the total infiltration flow to the sewers in dry weather (l/day), and E is the total dry weather industrial flow (l/day). DWF is defined as the flow in the sewer over 24 h when no rain has fallen for several days. Preliminary treatment includes flow measurement, flow separation, screening and comminution of solids, and removal of grit. The settlement of suspended solids in storm water should strictly be covered in primary settlement; however, the decision as to what is stormwater influences its treatment. This aspect is dealt with in this chapter.

2.2 Flow measurement

Measurement of flows is a major consideration in sewage treatment. In general, V notch and rectangular thin plate weirs are more suitable for effluent measurement or for intermittent use in influent channels whereas standing wave flumes are more suitable for continuous monitoring of raw sewage flows. Broad crested weirs find applications in separation of storm flows. The equations for weirs are empirical expressions and must be used within the conditions under which they were derived. A standard hydraulic text should be consulted [2] if in doubt; UK practice is adequately covered in the relevant standards [3].

2.2.1 V-notch weirs

The general expression for the flow Q (m^3/s) over a V-notch weir of the type shown in Fig. 2.3 is

$$Q = Ch^{2.5} \tag{2.3}$$

where h is the vertical head through the notch measured from the 'vertex' of the notch to the TWL, which in turn is measured a distance of 3 to 4 times the maximum head, upstream of the weir. C is a constant which includes a term for the velocity of approach and is numerically equal to

$$C = 1.4 \tan \frac{\theta}{2} \tag{2.4}$$

where θ is the notch angle.

$$\text{For } \theta = 90^\circ, \; Q = 1.4\, h^{2.5} \tag{2.5}$$

other common types are

$$\theta = 60^\circ, \; Q = 0.81\, h^{2.5} \tag{2.6}$$

or

$$\theta = 53^\circ, \; \text{i.e. notch width} \cong h$$

$$Q = 0.70\, h^{2.5} \tag{2.7}$$

The nappe must be fully aerated and the conditions should be as in Fig. 2.3 (formulae for other conditions are available [3]). A condition of Formulae (2.4) to (2.7) is that the approach velocity term $V^2/2g$ should be $\ll h$, where V is the mean approach velocity (m/s) and g is the gravitational acceleration (9.8 m/s^2).

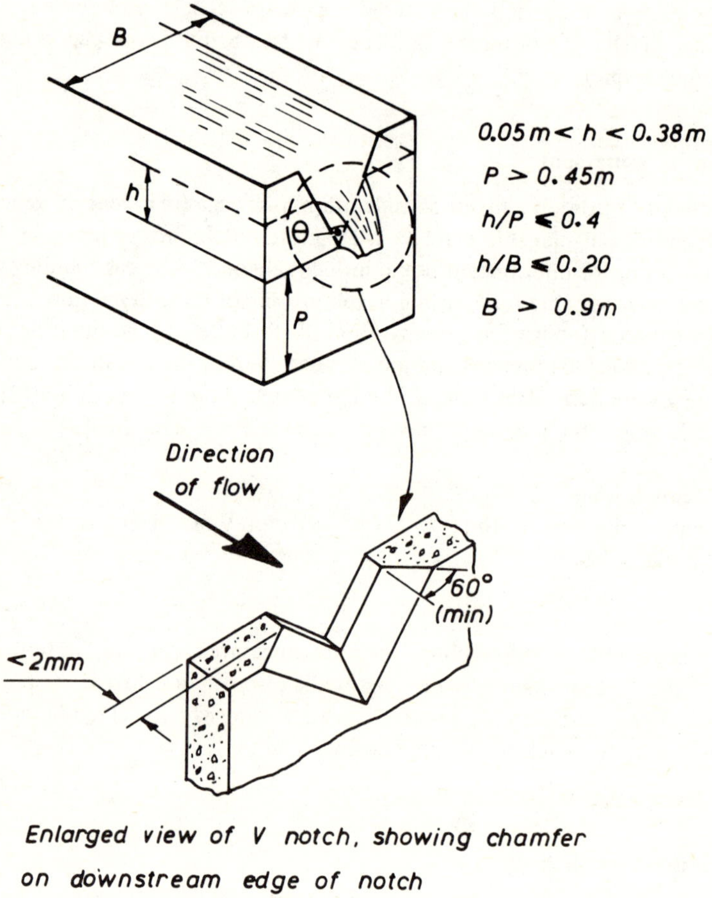

Fig. 2.3 V-notch weir details (by courtesy of British Standards Institution – See BSCP 3680 part 4*a*)

Example 2.1

Determine the flow range within which a 90° V-notch weir may be used according to BS 3680 for a 1 m wide channel.

Calculation

$$\frac{h}{B} \ngtr 0.2, \text{ i.e. } h \ngtr 0.2 \text{ m for } B = 1 \text{ m.}$$

$$\frac{h}{P} \ngtr 0.4, \text{ i.e. for the minimum value of } P \text{ of } 0.45 \text{ m;}$$

$$h \ngtr 0.18 \text{ m}$$

Thus by increasing P to 0.5 m the ratio of h/B becomes the limiting criterion for maximum flow.

A minimum value for h of 0.05 m is required.

Thus

$$Q_{\text{min}} = 1.4\,(0.05)^{2.5}$$
$$= 0.0008 \text{ m}^3/\text{s or } 0.8 \text{ l/s}$$

and

$$Q_{\text{max}} = 1.4\,(0.2)^{2.5}$$
$$= 0.025 \text{ m}^3/\text{s}$$
$$= 25 \text{ l/s}$$

□ □ □

It should be noted that under maximum flow conditions the upstream velocity is

$$\frac{Q_{\text{max}}}{B(P + h)} = \frac{0.025}{1(0.5 + 0.2)} = 0.036 \text{ m/s}$$

and an approach velocity term $V^2/(2g)$ would be 6.5×10^{-5} m which is negligible compared with a value for h of 0.2 m.

2.2.2 Rectangular weirs (thin plate)

(a) Full width or suppressed weirs

These run the full width of the channel and consequently the nappe is not fully aerated. BS 3680 suggests the use of the Rehbock equation although others exist [4], i.e.

$$Q = 2.95\,B\left(0.602 + 0.083\,\frac{h}{P}\right)(h + 0.0012)^{1.5} \tag{2.8}$$

The operational restrictions are that $h/P \leqslant 1.0$; $0.03 \text{ m} < h < 0.75 \text{ m}$; $B \geqslant 0.3$ m and $P \geqslant 0.1$ m.

Example 2.2

A full width thin plate rectangular weir is to be used in a 0.4 m wide channel, P being 0.20 m. Calculate the maximum allowable flow.

Calculation

Since $h/P \not> 1$, then $h \not> 0.20$ m (i.e. this is within the range 0.03 – 0.75 m.). Therefore

$$Q = 2.95 \times 0.4\,(0.602 + 0.083 \times 1)(0.20 + 0.0012)^{1.5}$$

$$= 0.073 \text{ m}^3/\text{s or } 73 \text{ l/s}$$

□□□

It should be noted that if P and B are increased, the allowable flow also increases. Also, the requirement that the full width weir be aerated (i.e. the nappe must not cling to the downstream side of the weir) presumes some ventilation holes or pipes downstream of the weir behind the waterfall and above the maximum downstream water level.

(*b*) *Partial width weirs*

A primary requirement is that the approach channel width shall be $> (4h + b)$ where h and b are as in Fig. 2.4. Under these conditions the contractions are fully developed and the expression relating discharge to head is

$$Q = 1.82\left(1 - 0.1\,\frac{h}{b}\right) bh^{3/2} \tag{2.9}$$

Other conditions are given in Fig. 2.4. The requirement that the walls be a distance of $2h$ from the sides of the weir produces very wide channels, as shown in the example below.

Example 2.3

Assuming a weir of minimum width 0.3 m, determine the head h and hence the channel width for a flow of 0.037 m^3/s.

$$Q = 0.037 = 1.82\,(0.3 - 0.1h)\,h^{3/2}$$

i.e.

$$0.0203 = (0.3 - 0.1h)\,h^{3/2}$$

Using an iterative solution, assume first that $0.1h$ is small thus

$$h \cong \left(\frac{0.0203}{0.3}\right)^{2/3} = 0.166$$

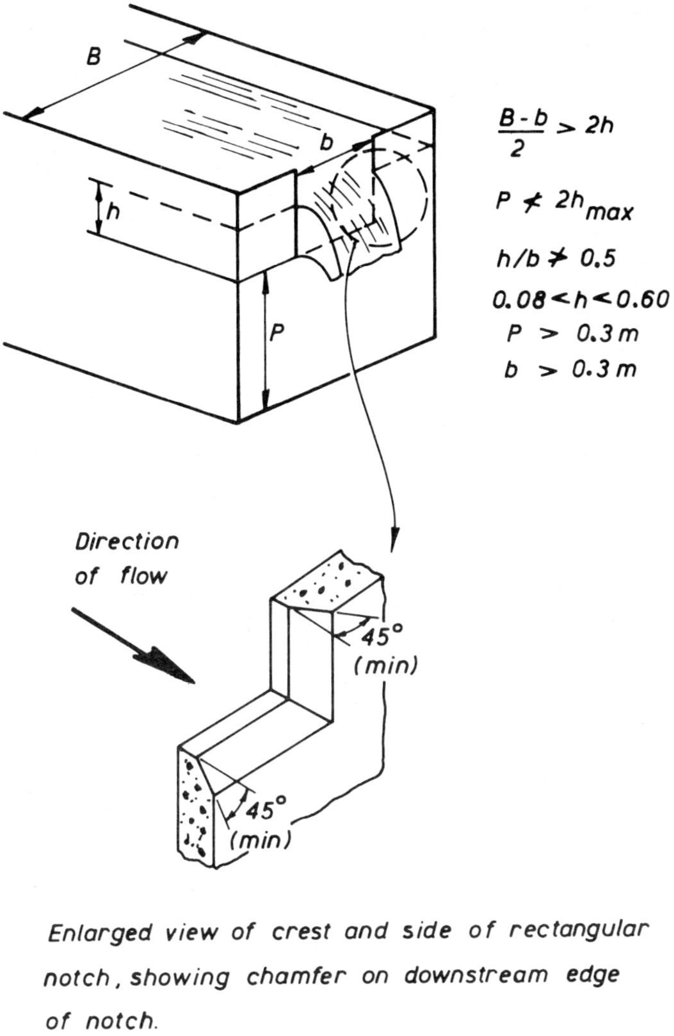

Fig. 2.4 Partial width rectangular weir (by courtesy of British Standards Institution – see BSCP 3680 part 4*a*)

using this in the expression $(0.3 - 0.1h)h^{3/2}$ gives 0.0192 and hence $Q = 0.035$. However, it is obvious that h/b exceeds 0.5, therefore the weir must be extended. Therefore, try $b = 0.4$, hence

$$\frac{0.037}{1.82} = (0.4 - 0.1h)h^{3/2}$$

and assuming $0.1h$ to be small

$$h = \left(\frac{0.037}{1.82 \times 0.4}\right)^{2/3} = 0.137 \text{ m}$$

Trying $h = 0.14$ m gives

$$1.82\,(0.4 - 0.1 \times 0.14)\,0.14^{1.5} = 0.037 \text{ m}^3/\text{s}$$

i.e. this is a correct solution.

The walls must be $2h$ away from the weir sides, thus $B = 4h + b$ or 0.96 m; compare this channel width with that in Example 2.2.

□ □ □

It should be noted that in BS 3680 p. 24§ g, a correction for h would be necessary if P is low, i.e. if $B(h + P) < 10bh$ the value of h in Equation (2.8) must be replaced by h^1 where

$$h^1 = h + 1.4\,\frac{V^2}{2g} \tag{2.10}$$

In this case $B(h + P) = 10bh$ gives a value for P of $(10 \times 0.14 \times 0.4)/0.96 - 0.14$ or 0.44 m, i.e. P is satisfactory if designed at 0.44 m.

2.2.3 Flumes

A common method of measuring flows containing gross solids (which would settle out if a weir were used), is a standing wave flume. Fig. 2.5 shows a typical case together with dimensional restraints. Reference 3 gives more detailed forms of the relationship, however for most purposes the expression below is satisfactory, i.e.

$$Q = 1.71\,bh^{3/2} \tag{2.11}$$

where b is the throat width and h is the depth of flow through the flume, measured a distance of $4h_{max}$ upstream. The above expression covers the case of a flume with side contractions only. Flow measurement is often carried out by placing a recorder attached to a float upstream of the flume. Only the single measurement h is required for a given value of b and this type of system lends itself readily to the use of a float actuated integrating flow meter situated in the upstream well.

The velocity requirement is that for a distance of $20B$ upstream of the flume, the flow should be subcritical. It is preferable therefore to ensure that for a rectangular channel the Froude number $[V/\sqrt{(gh)}]$ is less than 0.6. Thus for $h = 0.4$ m the approach velocity should be no greater than approximately 1.2 m/s. This ability of the structure to operate at high approach velocities is important in sewage treatment works where solids have to be kept in suspension.

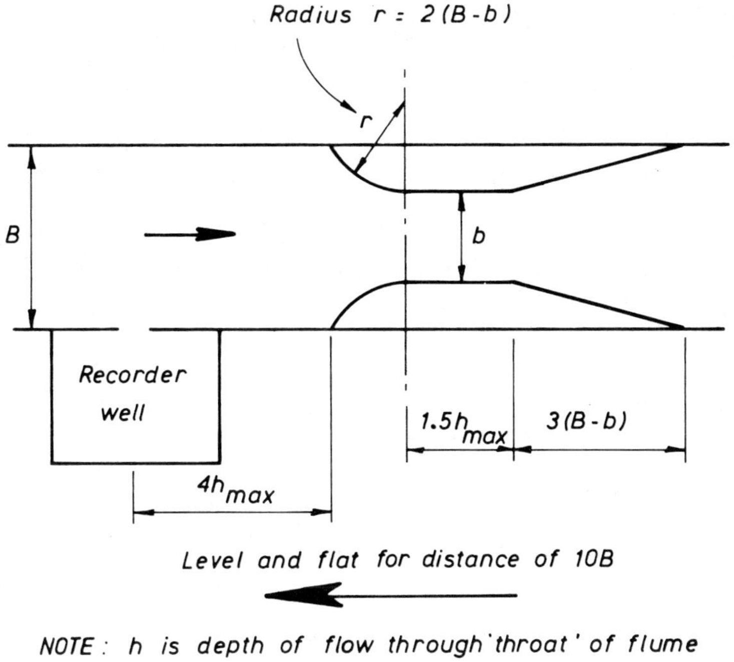

Fig. 2.5 Standing wave flume

The use of a standing wave flume to control upstream grit settlement channels is given in Section 2.5.3 and a further example of the use of such a flume as a regulatory device is given in Section 2.3.

2.2.4 Modular limits and head losses

For the standing wave flume discussed in the previous section it is clear that unsatisfactory downstream conditions (e.g. reducing slope of the channel invert or a flow restriction) could lead to the standing wave advancing through the tail of the flume and submerging or drowning the device. In order to avoid this condition, the device is operated at less than its modular limit. This is defined for the flume as

$$\text{Modular limit} = \frac{\text{Total energy head downstream of the standing wave measured with respect to the channel invert}}{\text{Total energy head upstream of the flume measured with respect to the channel invert}} \tag{2.12}$$

The denominator is $h + V^2/(2g)$ where V is the approach velocity and the numerator is the downstream depth of flow plus the term $V^2/(2g)$ where V is the downstream channel velocity. The modular limit for a standing wave flume is usually taken as 0.7 and, because the $V^2/(2g)$ terms are small, it is often satisfactory to design the downstream conditions such that the depth of flow downstream is $\ngtr (2/3)h$.

If a broad crested weir were used in a channel, the modular limit would be defined similar to Equation (2.12) but with the total energy heads being measured with respect to the weir crest.

For sharp crested or thin plate weirs, a satisfactory downstream flow level is usually defined in terms of the minimum acceptable head loss through the weir. For a partial width weir the head loss should be assumed at least equal to the total upstream head over the weir $[h + V^2/(2g)]$ plus 0.05 m. For a constant slope and cross-section channel containing a partial width thin plate weir, the downstream flow will be satisfactory provided the height of the weir crest, P is greater than 0.3 m and is also not less than twice the maximum depth of flow over the weir. Apart from the modular limits and head losses there are many other criteria involved in the selection of a flow measuring device and for further information on this topic the reader is referred to a specialist text [5].

2.3 Flow separation

In many instances some form of flow separation is required, e.g. in the UK, for a partially separate or combined sewerage system some method of separating flows greater than 3 DWF is required, the excess being transferred to stormwater settlement tanks. One method is to use side weirs that are controlled by a downstream standing wave flume as below.

Example 2.4

A sewage works is designed for a future population (N) of 10 000 persons with a daily water usage (G) of 230 l/cap, (the industrial effluent (E) and the infiltration (I) will be neglected in this example. Estimate the dimensions of a broad crested side weir (controlled downstream by a standing wave flume) used to separate flows between 3 DWF and maximum stormwater flow Q_{max} (6 DWF).

Calculation

$$3\ \text{DWF} = 3NG + I + 3E$$

$$= 6900\ \text{m}^3/\text{day}$$

$$= 0.08\ \text{m}^3/\text{s}$$

and

$$Q_{max} = NG + I + 1360N + 3E$$

$$= 15\,900\ \text{m}^3/\text{day}$$

$$= 0.18\ \text{m}^3/\text{s}$$

The difference is 0.10 m^3/s.

The formula for a broad crested weir is

$$Q = 1.71\ W_s h_s^{3/2} \tag{2.13}$$

where W_s is the side weir width (m) and h_s is the side weir head (m). For this calculation using a double sided weir

$$W_s = \frac{1}{2}\left(\frac{0.1}{1.71\ h_s^{3/2}}\right)$$

Ideally, the side weir should be capable of spilling the difference between Q_{max} and 3 DWF while keeping the same maximum head h_{max} through the flume, where h_{max} corresponds to 3 DWF. In practice the difference in head in the flume for flows upstream of Q_{max} and 3 DWF will vary by the maximum head over the side weir h_s, thus h should be kept low, e.g. 0.05 m.

Therefore

$$W_s = \frac{1}{2}\left(\frac{0.1}{1.71 \times 0.0112}\right) = 2.6 \text{ m each side}$$

If a width for the flume of say 0.3 m has previously been calculated, the crest of the side weir would be above the channel invert by the value of the flume head at 3 DWF, i.e.

$$h = \left(\frac{0.08}{1.71 \times 0.3}\right)^{2/3} = 0.290 \text{ m}$$

□ □ □

At maximum flow the flume would actually be operating with a head of 0.290 + 0.05 m = 0.340 m; equivalent to a flow of 0.3 x 1.71 x $(0.340)^{3/2}$ or 0.10 m^3/s, i.e. 3.8 DWF. Increasing the flow through the works to 3.8 DWF would greatly increase the primary and secondary treatment costs. Therefore three alternatives are possible:

(1) reduce the maximum head over the side weir to 0.025 m, which will increase W_s to (1/2)[0.1/(1.71 x 0.004)] m or 7.4 m and reduce the maximum head through the flume to 0.315 m and the corresponding flow to 0.3 x $1.71(0.315)^{3/2}$ m^3/s or 0.091 m^3/s or 3.4 DWF.

(2) keeping the original h_s value of 0.05 m, site the crest of the side weir at 0.290 – (0.05/2) m or 0.265 m above the channel invert. The weir then begins to spill when the head through the flume is h = 0.265 m and is discharging its maximum flow when h = 0.315 m, i.e. flows of 0.070 m^3/s (2.6 DWF) and 0.091 m^3/s (3.4 DWF), respectively.

(3) use a motorized penstock as a side weir driven by feedback from a sensor placed just upstream of the flume.

2.4 Stormwater settlement

It has been common UK practice to allow $6h$ capacity at DWF for stormwater settlement, i.e. if 3 DWF is being discharged to storm tanks, the nominal retention is approximately $2h$. On the suggestion of the technical committee on storm overflows [1], the maximum flow to the works from a combined sewerage system would be limited (by an overflow device) to 6 DWF.

$$6\ \text{DWF} = Q_{max} = NG + I + 1360N + 3E \tag{2.2}$$

where NG, I and E are defined as in the introduction to this chapter. 3 DWF would be

$$3\ \text{DWF} = 3NG + I + 3E \ \ \text{l/day} \tag{2.1}$$

Stormwater treatment is necessary for the difference, i.e.

$$N(1360 - 2G)\ \ \text{l/day} \tag{2.14}$$

The report quoted above suggested a volume in the settlement tank of 68 l/cap. instead of using $6h$ capacity at DWF. This would give a retention period of

$$\frac{68N24}{N(1360 - 2G)} \text{ or } \frac{1632}{(1360 - 2G)}\ h$$

If G is 230 l/cap . day, the retention time is therefore 1.8 h. Note that the above definition of Q_{max} (i.e. the old 6 DWF) is reduced only slightly if G is reduced whereas the 3 DWF term is reduced by twice as much. Therefore the term $N(1360 - 2G)$ increases as G diminishes. The retention time in the settlement tank is therefore reduced, e.g. at 160 l/cap . day the retention time is

$$\frac{68 \times 24}{1360 - 2 \times 160}\ h = 1.57\ h$$

Usually several tanks are used filling sequentially but overflowing simultaneously when full. At many works the first tank has no overflow and is used to store the first flush of the storm. The settled sewage/stormwater is then returned to the works inlet after the storm has passed.

2.5 Grit removal

2.5.1 Settlement velocities

Consider a spherical, discrete (non-flocculating) particle of density ρ_p(kg/m^3) acted upon by gravity g(m/s^2) and a drag force in a fluid of density ρ_w(kg/m^3). Then

$$mg - \frac{m}{\rho_p} g\rho_w = C_D \rho_w A \frac{u^2}{2} \tag{2.15}$$

where u is the particle's terminal velocity (m/s); A its projected area (m/s); C_D is a drag force coefficient and m the particle mass (kg). Thus

$$u^2 = \frac{2g}{C_D}\left(\frac{\rho_p - \rho_w}{\rho_w}\right) \times \left(\frac{\text{volume of particle}}{\text{area of particle}}\right) \tag{2.16}$$

and for a sphere

$$u = \left(\frac{4}{3}\frac{gd}{C_D}(R_D - 1)\right)^{1/2} \tag{2.17}$$

where R_D is the relative density of the particle (specific gravity), and d is the particle diameter (m). C_D is a function of the Reynolds number R_e

$$R_e = \frac{ud}{\upsilon} \tag{2.18}$$

where υ is the kinematic viscosity of the fluid (m^2/s).

For values of $R_e < 0.5$; $C_D \cong 24/R_e$ and

$$u = \frac{gd^2}{18\upsilon}(R_D - 1) \tag{2.19}$$

For values of $R_e > 10^4$; $C_D \cong 0.4$ for spheres and therefore

$$u = \left(\frac{gd}{0.3}(R_D - 1)\right)^{1/2} \tag{2.20}$$

Unfortunately, for most cases of grit settlement, in practice $0.5 < R_e < 10^4$. For this region C_D is found from the expression

$$C_D = \frac{24}{R_e} + \frac{3}{\sqrt{R_e}} + 0.34 \tag{2.21}$$

Iterative (repeated) calculations are therefore necessary to determine the settlement velocity of grit particles.

Example 2.5

Two particles of density $\rho_p = 2600$ kg/m^3 and diameters 0.1 and 3 mm, respectively, settle through water having kinematic viscosity $\upsilon = 1.3 \times 10^{-6}$ m^2/s and a density of 1000 kg/m^3. Determine the settlement velocities of the two particles.

Calculation

Consider the particle $d = 0.1 \times 10^{-3}$ m and guess a velocity of 1 m/s (a most unlikely high value in practice). Then

$$R_e = \frac{1 \times 0.1 \times 10^{-3}}{1.3 \times 10^{-6}} = 76.9$$

and

$$C_D = \frac{24}{76.9} + \frac{3}{\sqrt{76.9}} + 0.34 = 0.99$$

and

$$u = \left(\frac{4}{3}\frac{0.1 \times 10^{-3}}{0.99} \times 9.81 \times 1.6\right)^{1/2} = 0.046 \text{ m/s}$$

Repeating the steps using this value of u enables a new value of R_e to be found, i.e.

$$R_e = \frac{0.046 \times 0.1 \times 10^{-3}}{1.3 \times 10^{-6}} = 3.54$$

and hence a new value of C_D is found which in turn leads to a further iteration. A halt is usually called in this process when subsequent iterations differ by 2% or less. In this case the values for u tend towards 6.3×10^{-3} m/s. The settlement velocity of the 3 mm particle is approximately 0.37 m/s by a similar system.

□ □ □

2.5.2 The design particle

Since sewage contains a wide range of grit particles, it is necessary to specify, for design purposes, the smallest size of particle which it is hoped to remove completely. Traditionally this has been taken as the particle with a settlement velocity of approximately 0.03 m/s. Most grit removal processes attempt to allow such particles to settle while keeping a sufficiently high forward velocity through the grit channel or grit trap such that organic material is kept in suspension. Therefore, in the simplest form of grit removal system, i.e. a long constant velocity channel, a forward velocity of 0.30 m/s is used. For a depth of flow H(m), a design particle will just settle within the channel if the ratio of H to length L(m) is 0.03/0.3, i.e. $L = 10H$.

In practice, owing to turbulence at inlet and outlet, values of L of as great as $25H$ are used.

2.5.3 Grit channels

These are effectively long channels in which the grit settles and the organics remain suspended.

From the Manning formula, the forward velocity in a channel of fixed slope S is given by the expression

$$V = \frac{r^{2/3} S^{1/2}}{n} = \text{constant} \left(\frac{A}{P}\right)^{2/3} \tag{2.22}$$

where V, r, S and n, A and P are defined in Chapter 1, Section 1.3 and in the nomenclature to Chapter 1. For a square section channel, for given values of S and n, the forward velocity at maximum flow

$$V_{max} \propto \left(\frac{H}{3}\right)^{2/3} \tag{2.23}$$

and for half maximum flow is

$$V_{1/2\,max} \propto \left(\frac{H}{4}\right)^{2/3} \tag{2.24}$$

Such variations in V clearly make it impossible to prevent scouring of grit at high flows or prevent organic deposition of organics at low flows. However, if a control (standing wave flume) is placed downstream of the grit channel, the depth of flow in the grit channel H is the same as that in the flume h, and the discharge is

$$Q = 1.71\, bH^{3/2} \tag{2.11}$$

Thus for the flume

$$\frac{dQ}{dH} = \text{constant}\, H^{1/2} \tag{2.25}$$

Upstream of the flume the channel discharge is given by

$$Q = VA \tag{2.26}$$

For constant V

$$\frac{dQ}{dH} = V \frac{dA}{dH} \tag{2.27}$$

Thus

$$\frac{dA}{dH} = \text{constant}\, H^{1/2} \tag{2.28}$$

and since

$$\frac{dA}{dH} \cong W \tag{2.29}$$

where W is the channel width (m) at the water surface, then for a grit channel (controlled downstream by a standing wave flume) in which $W = H^{1/2}$ the velocity remains constant for all depths of flow.

This is the equation of a parabola and for such a shape the area of flow is approximately two-thirds the area of the rectangle formed from the product of $H \times W$ (at maximum H). Thus

$$Q_{max} = V\,0.667\,WH \tag{2.30}$$

If V is given the value of 0.305 m/s (the old 1 ft/s value), then W at maximum H is

$$W = 4.92\,\frac{Q_{max}}{H_{max}} \tag{2.31}$$

Because parabolic sections are difficult to construct, a trapezoidal section channel is usually fitted to the theoretical curve, as in Example 2.6.

Example 2.6

An outfall sewer of 900 mm diameter discharges a maximum of 0.82 m^3/s. Calculate the dimensions of the grit channels.

Calculation

For the flume $Q = 1.71\,bH^{3/2}$ and b is usually $2W/3$ thus $Q = 1.14\,WH^{3/2}$.

However, at maximum flow the flume must not surcharge the upstream outfall sewer. Thus, assume $H_{max} \cong 0.9d$ where d is the inlet sewer diameter. Then

$$Q_{max} = 1.14\,W(0.9d)^{1.5}$$

and for the above values of Q_{max} and d of 0.82 m^3/s and 900 mm

$$W = \frac{0.82}{1.14(0.81)^{1.5}} = 0.99 \text{ m or approximately 1 m}$$

The standing wave flume throat width is therefore 0.67 m. The maximum depth of flow in the grit channel is therefore

$$H_{max} = \left(\frac{0.82}{1.71 \times 0.67}\right)^{2/3} = 0.80 \text{ m}$$

The width of the channel at the water surface in the parabolic channel at maximum flow is

$$W = \frac{4.92 \times 0.82}{0.80} = 5.04 \text{ m}$$

The total grit channel width at various flows can be calculated as in Table 2.1

Table 2.1 Discharge, depth of flow and channel width

(a) Discharge $Q(m^3/s)$	(b) Corresponding flume depth H(m)	(c) Grit channel width (m)	(d) $W/5$ (m)
0.7	0.72	4.78	0.96
0.6	0.65	4.54	0.91
0.5	0.58	4.24	0.85
0.4	0.50	3.94	0.79
0.3	0.41	3.60	0.72
0.2	0.31	3.17	0.63
0.1	0.20	2.46	0.49
0.05	0.12	2.05	0.41

and the shape of the grit channel theoretically defined, e.g. for a flow of 0.5 m^3/s

$$H = \left(\frac{0.5}{1.71 \times 0.67}\right)^{2/3} = 0.58 \text{ m}$$

The width of the grit channel for this flow is

$$W = \frac{4.92 \times 0.5}{0.58} = 4.24 \text{ m}$$

A single channel would possess too large a radius of curvature of the invert and would make channel cleaning difficult. A choice of 5 channels would appear reasonable, each one having widths as in column (*d*) in Table 2.1 for given depths of flow. Fig. 2.6 shows the plot of the theoretical parabola and a trapezoid fitted to the curve. Where the two curves are conjoint the velocity is close to the ideal 0.305 m/s. However, for points A, B and C the fitted trapezoid is significantly in error. To calculate the error consider:

Point B

W(real) = 4.6 m

$H = 0.55$ m

$$Q = 1.71 \times 0.67 \times (0.55)^{1.5}$$
$$= 0.47 \text{ m}^3/\text{s}$$

Total area of flow in five channels $\cong$ 1.51 m^2
Therefore $V_B = Q/A$

$$V_B = \frac{0.47}{1.51} = 0.311 \text{ m/s}$$

Point A

W(real) = 4.6 m

$H = 0.80$ m

$$Q = 1.71 \times 0.67 \times (0.80)^{1.5}$$
$$= 0.82 \text{ m}^3/\text{s}$$

Σ area = 2.66 m^2

$$V_A = \frac{0.82}{2.66} = 0.308 \text{ m/s}$$

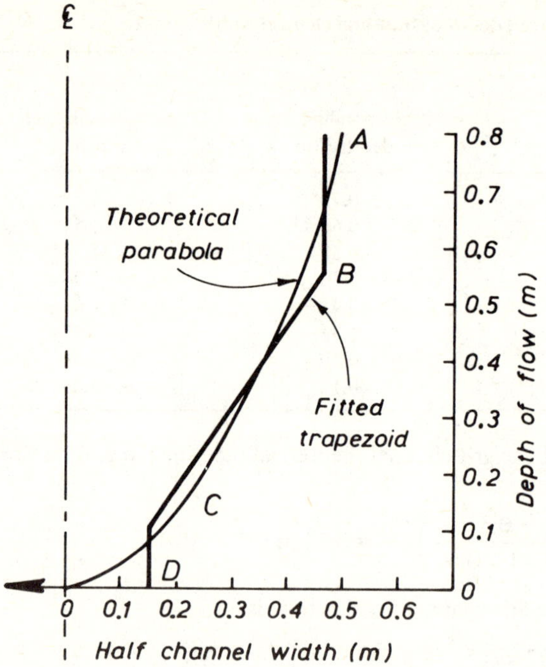

Fig. 2.6 Fitting trapezoid section to theoretical grit channel parabola

Both are satisfactorily close to the ideal value of 0.305 m/s. In practice each channel would have its own control, however the width would be one-fifth that of the single flume. Thus

$$H = \left(\frac{Q/5}{1.71 \times 0.67\ W/5}\right)^{1.5}$$

is unchanged.

□ □ □

When low flows occur one or more channels are taken out of commission; the discrepancy between the theoretical curve and the fitted curve at *C* is therefore not critical. N.B. A choice of slope for the trapezoid of 1.75 to 1 is often satisfactory with the top third of the channel sides being vertical.

2.5.4 Scouring velocities

In practice grit channels operate by using the different scouring rates of grit and the lighter organic materials. It is sometimes useful to calculate the velocity at which different materials will remain in suspension; once particles have been deposited, the system becomes extremely complex. The following discussion shows some of the more important aspects of scouring theory.

The scouring of grit may be described by the Shields[6] formula

$$V = \sqrt{\left(\frac{8K}{f} g(R_D - 1)d\right)} \tag{2.32}$$

where V is the velocity at which the particle will begin to scour (m/s) and K is a constant (which varies with Reynolds number) between 0.03 to 0.06; f is a friction factor (0.03 is a common value); g is the gravitational acceleration (m/s^2); R_D is the particle relative density (specific gravity) and d is the particle diameter.

Example 2.7

Assuming a mean velocity of 0.305 m/s determine the size of spherical grit particle ($R_D = 2.65$) and spherical organic material ($R_D = 1.05$) which will remain in suspension in a grit channel for which $K = 0.05$ and $f = 0.03$.

Calculation

Rearranging Equation (2.27) and assuming a velocity of 0.305 m/s

$$d = \frac{V^2 f}{8Kg(R_D - 1)} \tag{2.33}$$

For grit

$$d = (0.305)^2 \frac{0.03}{8 \times 0.05 \times 9.81\ (1.65)} = 0.43 \times 10^{-3}\ \text{m}$$

for the organics

$$d = (0.305)^2 \frac{0.03}{8 \times 0.05 \times 9.81\ (0.05)} = 14 \times 10^{-3}\ \text{m}$$

i.e. it would be expected that grit particles less than 0.43 mm diameter and organics of relative density 1.05 and less than 14 mm diameter will remain in suspension.

□ □ □

It should be noted that many organics will have R_D values closer to unity. In practice, once a particle has settled to the bed in a non uniform sediment it may require a greater velocity to scour the particle than would be evident from the above calculations since larger particles on the bed can influence the shear forces and 'protect' the finer material.

Questions

Question 2.1

A 600 mm diameter sewer discharges 0.30 m^3/s at maximum flow to a channel and hence to two identical parabolic grit channels which are controlled by a standing wave flume whose throat width is 2/3 the inlet channel width. The inlet channel TWL has a maximum depth of flow of 0.54 m. Calculate the width of each grit channel at maximum flow and 1/3 maximum flow.

Question 2.2

A flume is used to settle out a useful ore (R_D = 3.7) while scouring a crushed rock waste (R_D = 2.7). All particles are spherical and f and k in the Shields formula are 0.03 and 0.05, respectively. Calculate the minimum size of useful ore that can be settled out without being contaminated by the rock waste which has a maximum diameter of 3 mm.

References

1 Ministry of Housing and Local Government (1963), *Technical committee on storm overflows and the disposal of storm sewage; Interim report*, HMSO, London.
2* King, H. W. and Brater, E. F. (1963), *Handbook of hydraulics*, McGraw-Hill.
3* BS 3680 Part 4A (1965), *Measurement of liquid flow in open channels: Part 4, Weirs and Venturi Flumes,* British Standards Institution, London.
4 Degremont, G. (1973), *Water Treatment Handbook* (English translation), 4th Edn, Rueil-Malmaison, France.
5 Bos, M. G. (1976), *Discharge Measurement Structures*, Report no. 161, Delft Hydraulics Laboratory, Delft.
6 Shields, A. (1963). An wendung der aenlichkeitsmechanik und der turbulenz-forschung an die geschiebebegung. Mitt. der Preuss. *Versuchsanstalt fur Wasserban und Schiffban*, **26**, Berlin.

* Suggested further reading.

3
Settlement of solids

Nomenclature

Symbol	*Definition and units*	*Dimensions*
Q	Rate of flow into tank, m^3/s, m^3/day, m^3/h, l/s etc. as stated	L^3T^{-1}
V_0	Design settlement velocity of tank. For horizontal flow tanks this is the flow/surface area (see Fig. 3.1), m/s, mm/h, etc.	LT^{-1}
V	Mean forward velocity through tank; later sludge bulk settling velocity, m/s, m/h as stated	LT^{-1}
B	Width of individual tank, m	L
H	Depth of tank, m	L
L	Length of tank, m	L
υ	Kinematic viscosity of fluid, m^2/s	L^2T^{-1}
H_x	Height above invert of particle with settlement velocity V_x, m	L
V_x	Settlement velocity of particle, m/s or mm/s	LT^{-1}
R_D	Relative density (specific gravity) of particle, dimensionless	–
g	Gravitational constant, m/s^2	LT^{-2}
d	Particle diameter, m	L
C	Solids concentration, kg/m^3, mg/l etc.	ML^{-3}
r_b	Radius of base of a cone or hopper, m	L
h	Depth or height of cone, m	L
N	Number of segments dx in Fig. 3.3	–
n	Number of tanks	–
T_R	Time ratio; degree of ideality of tank behaviour	–
B_0	Total breadth of tanks, i.e. $B_0 = nB$, m	L
E	Efficiency of removal of SS (fraction)	–
S_i	Inlet suspended solids concentration, mg/l	ML^{-3}
SVI	Sludge volume index see Equation (3.14)	–
SDI	Sludge density index = 100/SVI	–

Q_e	Effluent flow rate from settlement tank, m^3/s, m^3/day etc. as stated	L^3T^{-1}
q	Returned (settled) sludge flow, m^3/s, m^3/day etc. as stated	L^3T^{-1}
C_0	Initial concentration of SS later MLSS concentration, kg/m^3, mg/l	ML^{-3}
t	Time of settlement, s, h, day	T
Z	Depth in settling column, m	L
$Z_i t_i C_i$	As above parameters but relating to a time t_i	
SS	Suspended solids, usually taken as those solids retained on GF/C filter paper. As a dry mass/volume, kg/m^3, mg/l	ML^{-3}
x_0	Fraction of particles with velocities $\leqslant V_0$, dimensionless	–
x_T	Total settleable fraction, dimensionless	–
A	Cross-sectional or surface area of tank, m^2	L^2
D	Depth of settlement tank, minimum depth if tank slopes, m	L
WOR	Weir overflow rate, m^3/m day	L^2T^{-1}
C_u	Underflow solids concentration, kg/m^3, mg/l	ML^{-3}
C_e	Effluent solids concentration, kg/m^3, mg/l	ML^{-3}
V_s	Settlement velocity of sludge mass, m/h	LT^{-1}
Ψ	Solids flux through area of thickener, kg/m^2h	$ML^{-2}T^{-1}$
$\psi_j C_j V_j$	Solids flux, concentration and velocity at section j	
U	Underflow velocity or draw-down velocity q/A m/h	LT^{-1}
ψ_L	Limiting flux, i.e. the maximum rate at which the thickener can be loaded to avoid eventual solids overflow, kg/m^2h	$ML^{-2}T^{-1}$
ψ_I	Inlet solids loading expressed as a flux, kg/m^2h	$ML^{-2}T^{-1}$
V_I	Settling velocity of sludge mass on entry to tank at concentration C_0, m/s	LT^{-1}

3.1 Introduction

In Chapter 2 the removal of grit particles was discussed and equations relating settlement velocity to particle size and liquid type were presented. However, there is usually a large fraction of organic material suspended in domestic wastewaters which does not fully obey these simple relationships since it tends to coalesce or flocculate. This material is often removed in settlement tanks prior to secondary treatment (invariably so if conventional stone media percolating filters are used, see Chapter 4.) In addition, material of 70–80% organic content is produced by the secondary processes and usually must be removed prior to effluent discharge. Settlement is again a common and economical method of removal, and, in the case of the activated sludge process (see Chapter 4) is an inherent part of the secondary treatment scheme.

The wide range of suspension types found in wastewater treatment makes the use of unified theories of settlement very difficult. The following discussion first considers suspensions exhibiting behaviour which can be described by the equations for spherical particles given in Chapter 2, which includes primary settlement of domestic sewage. Secondly it shows how the behaviour of more strongly flocculating material may be described, e.g. primary settlement of domestic sewage treated with a flocculating agent such as aluminium chlorohydrate, and thirdly it describes the particular behaviour of activated sludge.

3.2 Overflow rates

Consider a discrete particle which reaches a terminal settlement velocity V_0 when settling in an ideal *horizontal* tank such as in Fig. 3.1 line '*a*'. Scouring is assumed not to occur, and particles which settle such that they just reach the invert prior to the end of the tank will be removed.

Now

$$Q = VBH \tag{3.1}$$

where Q is the forward flow through the tank; B and H are the breadth and depth of the tank, respectively, and V is the mean forward velocity, assumed to be constant at all points in the tank. By geometry

$$\frac{V}{V_0} = \frac{L}{H} \tag{3.2}$$

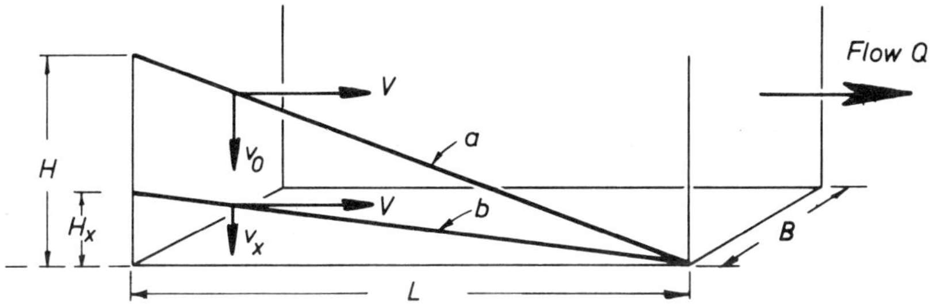

Fig. 3.1 Ideal settlement of discrete particles

where L is the length of the tank, and hence,

$$V_0 = \frac{Q}{LB} \tag{3.3}$$

i.e. the settlement velocity V_0 of particles 100% of which will just be settled out in the tank, is equal to the flow per unit *surface* area of the tank. This settlement velocity is referred to as the design settlement velocity of the tank and can be used to predict settlement efficiencies within the tank. The term Q/LB is also known as the surface loading rate or overflow rate of the tank.

Example 3.1
A flow of 20 l/s of a mixed suspension of discrete particles of relative density (specific gravity) of 2.6 is to be settled in an ideal settlement tank having a surface area of 5 m^2. Assuming Equation (2.19) applies ($\upsilon = 1 \times 10^{-6}$ m^2/s), calculate the diameter of particles, 100% of which will be settled. Also determine the effect of halving the depth.

Calculation

$$V_0 = \frac{Q}{LB} = \frac{20 \times 10^{-3}}{5} = 4 \times 10^{-3} \text{ m/s}$$

Therefore

$$d = \left[\frac{(V_0\, 18\upsilon)}{g(R_D - 1)}\right]^{1/2} = \left[\frac{4 \times 10^{-3} \times 18 \times 1 \times 10^{-6}}{9.81 \times 1.6}\right]^{1/2}$$

$$= 6.8 \times 10^{-5} \text{ m or } 0.068 \text{ mm}$$

Particles with relative densities of 2.6 with diameters greater than 0.068 mm will settle completely, some smaller particles will pass through the tank.

If the depth is halved, the forward velocity V is doubled. However the particles have only half the original vertical distance to fall.
Thus

$$\frac{2V}{V_0} = \frac{L}{H/2}$$

or

$$\frac{V}{V_0} = \frac{L}{H}$$

as before. Therefore, changing the depth has no effect unless extreme conditions of shear are produced, i.e. the tank ceases to be an ideal tank.

□□□

Fig. 3.1, line *b*, represents the path taken by a particle with settlement velocity V_x which enters the tank at height H_x in a homogeneous column of suspension of depth H. If it just settles out in a tank of design settlement velocity V_0 (i.e. $V/V_0 = L/H$), then by geometry

$$\frac{H_x}{L} = \frac{V_x}{V} \tag{3.5}$$

Substituting for $V = (L/H)V_0$

$$\frac{H_x}{H} = \frac{V_x}{V_0} \tag{3.6}$$

H_x/H is clearly the fraction of particles with velocity V_x which will be settled within the tank. The ratio of the settlement velocity of the particular particle to the design settlement velocity, gives the fraction settled.

Example 3.2
For the conditions given in Example 3.1, determine the proportion of particles with $R_D = 2.6$ and diameter 0.04 mm which will settle in the basin.

Calculation

$$V_x = \frac{gd^2}{18\upsilon}(R_D - 1) = \frac{9.81 \times (4 \times 10^{-5})^2(1.6)}{18 \times 1 \times 10^{-6}} = 0.0014 \text{ m/s}$$

Therefore the proportion settling is

$$\frac{V_x}{V_0} = \frac{1.4 \times 10^{-3}}{4 \times 10^{-3}} = 0.35 \text{ or } 35\%$$

□□□

3.3 Batch settlement analyses

Determination of V_x values for a mixed suspension is tedious and involves density and particle size analyses; it is common therefore to use a batch settlement analysis as follows. A column of the type shown in Fig. 3.2 is filled with the heterogenous suspension, no particle of which has a velocity great enough to settle a distance $> z_1$ prior to the initial sampling. Small samples (insignificant in terms of the total volume) are taken at each outlet at set times and the suspended solids concentration in the samples is determined. If the original concentration in the column is C_0 at time $t_0 = 0$ and some time later is C_i at time $t = t_i$ in a sample taken from depth z_i, then $C_0 - C_i$ of the original suspension has settlement velocities $\geqslant z_i/(t_i - t_0)$. When repeated for a range of

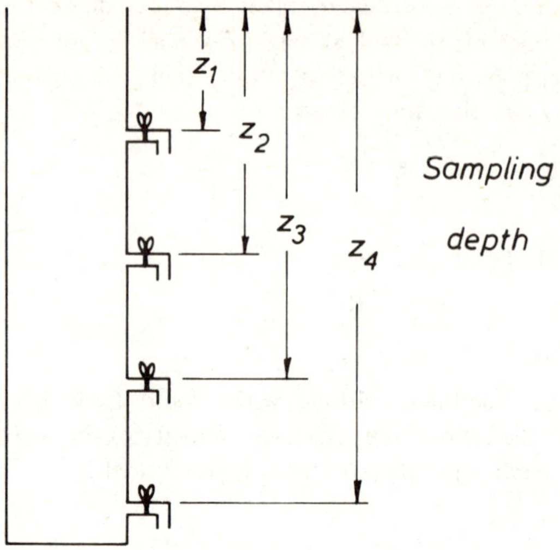

Fig. 3.2 Batch settlement container

z_i and t_i a series of values for C_i is obtained and a cumulative curve of proportion of particles with velocity $\leqslant$ the abcissa value can be plotted.

Example 3.3

The results of a batch settlement analysis carried out on a mixed suspension are given in Table 3.1. Plot the cumulative velocity profile of the suspended solids.

Table 3.1 Results of a batch settlement analysis

Number	Sample depth (m)	Sample time (h)	SS in sample (mg/l)
1, 2, 3	1, 2, 3	0	222 (averaged)
4	1	1	140
5	1	3	108
6	1	6	80
7	2	1	142
8	2	3	110
9	2	6	106
10	3	1	142
11	3	3	130
12	3	6	124
13	4	1	147
14	4	3	126
15	4	6	114

Calculation

Consider depth 2 m at time 3h. The velocities of particles found in the sample are $\leqslant 2000/(3 \times 3600)$ mm/s, i.e. 0.19 m/s. Therefore $(110/222) \times 100\%$ of particles have a settlement velocity < 0.19 mm/s. By a similar process we can construct Table 3.2.

Table 3.2 Fraction of particles remaining in sample and the settlement velocity of that fraction

Number	Velocity (mm/s)	SS remaining (fraction)	Number	Velocity (mm/s)	SS remaining (fraction)
4	0.278	0.63	10	0.833	0.64
5	0.093	0.49	11	0.278	0.59
6	0.046	0.36	12	0.139	0.56
7	0.556	0.64	13	1.11	0.66
8	0.185	0.50	14	0.370	0.57
9	0.093	0.48	15	0.185	0.51

Since the *fraction of particles* remaining in a sample is the same as the fraction of particles with velocity less than or equal to the value of the velocity pertaining to that sample, the values in Table 3.2 may be used to construct the cumulative settlement velocity curve shown in Fig. 3.3.

□ □ □

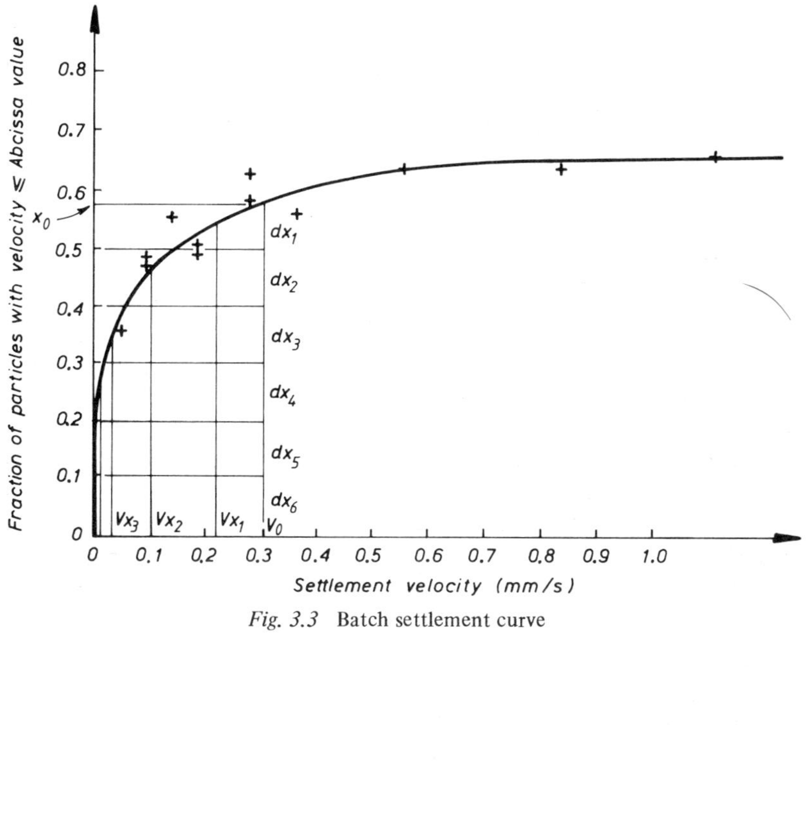

Fig. 3.3 Batch settlement curve

The next step is to determine two fractions of particles, (*a*) those whose velocity is greater than a chosen design settlement velocity V_0 and consequently will settle out, and (*b*) the fraction of those particles, which, although possessing velocities less than a chosen design velocity, V_0, will nonetheless contribute to the settled solids in the ratio V_x/V_0.

By inserting a chosen value for V_0 onto the abcissa in Fig. 3.3, a weight fraction can be determined $(1 - x_0)$ which has settlement velocities $> V_0$, i.e. 100% of such particles will settle in a tank designed such that $V_0 = Q$/surface area. For the weight fraction x_0 (i.e. whose settlement velocities V_x are $\leqslant V_0$) the proportion which will settle (x_s) is given by

$$x_s = \int_0^{x_0} \frac{V_x}{V_0}\, \mathrm{d}x \tag{3.7}$$

The value $\int_0^{x_0} V_x\, \mathrm{d}x$ is the area bounded by the ordinate (vertical axis) and the curve between 0 and x_0. For a chosen value of V_0 the total settleable solids x_T in a tank with design settlement velocity V_0 is

$$x_T = 1 - x_0 + \int_0^{x_0} \frac{V_x}{V_0}\, \mathrm{d}x \tag{3.8}$$

or

$$x_T = 1 - x_0 + \frac{1}{V_0} \sum_{i=1}^{N} V_{x_i}\, \mathrm{d}x_i \tag{3.9}$$

where N is the total number of segments. Choosing different values of V_0 gives a series of values of $1 - x_0$ and $\int_0^{x_0} V_x\, \mathrm{d}x$. The result of such an analysis can be shown in the following example.

Example 3.4
For design settlement velocities (V_0) in the range 0.2 to 0.8 mm/s find the predicted percentage settlement of particles for the previous batch settlement test.

Calculation
Inserting a value of $V_0 = 0.3$ mm/s on Fig. 3.3 shows that a fraction $x_0 = 0.58$ of particles have a velocity $\leqslant V_0$. Thus $1 - x_0$ in Equation (3.8) is $1 - 0.58$ or 0.42. There are several ways of evaluating the area between 0 and x_0 and the curve. The technique used by the author of splitting the region $0 - x_0$ into segments $\mathrm{d}x_i$ and assigning an average velocity V_{x_i} to each segment and summing the products $\mathrm{d}x_i \times V_{x_i}$ is relatively quick. Table 3.3 shows the method.

Table 3.3 Summation of $\frac{1}{V_0}\sum_{i=1}^{N} V_{x_i}\,dx_i$ given $V_0 = 0.33$

dx_i region	dx_i value	V_{x_i}	$dx_i V_{x_i}$
0.58–0.50	0.08	0.22	0.0176
0.50–0.40	0.10	0.10	0.010
0.40–0.30	0.10	0.04	0.004
0.30–0.20	0.10	0.015	0.0015
0.20–0.10	0.10	< 0.005 } neglected	
0.10–0.00	0.10	< 0.005 } neglected	
		$\Sigma\, dx_i V_{x_i} = 0.0331$	

Thus

$$\frac{dx_i V_{x_i}}{V_0} \cong \int_0^{x_0} \frac{V_x}{V_0}\,dx = \frac{0.033}{0.3} = 0.11$$

The total predicted settlement is given by

$$x_T = 1 - x_0 + \sum \frac{\Delta x_i V_{x_i}}{V_0}$$

$$= 0.53 \text{ or } 53\%$$

□ □ □

Similarly, by choosing other values of V_0, Fig. 3.4 may be constructed. In practice, the overflow rates at maximum flow are often 0.34 to 0.68 mm/s (4 ft/h to 8 ft/h). For the sample described above, little theoretical removal efficiency is gained by using the lower value of 0.34 mm/s.

3.3.1 Upward flow tanks

The foregoing discussion is applicable to horizontal flow tanks. If a particle exists in an ideal upward flow tank with cross-sectional area A and overflow rate Q then the liquid upward flow velocity has a magnitude Q/A or V_0 (the design settlement velocity) as before. The resultant of V_0 acting vertically upwards and the settlement velocity of the particle V_x will be upward if $V_x < V_0$; zero if $V_x = V_0$ and downwards if $V_x > V_0$. The total removal for a suspension similar to that in Fig. 3.3 would be

$$x_T = 1 - x_0 \qquad (3.10)$$

i.e. there will be no contributions by particles having a velocity $V_x \leqslant V_0$.

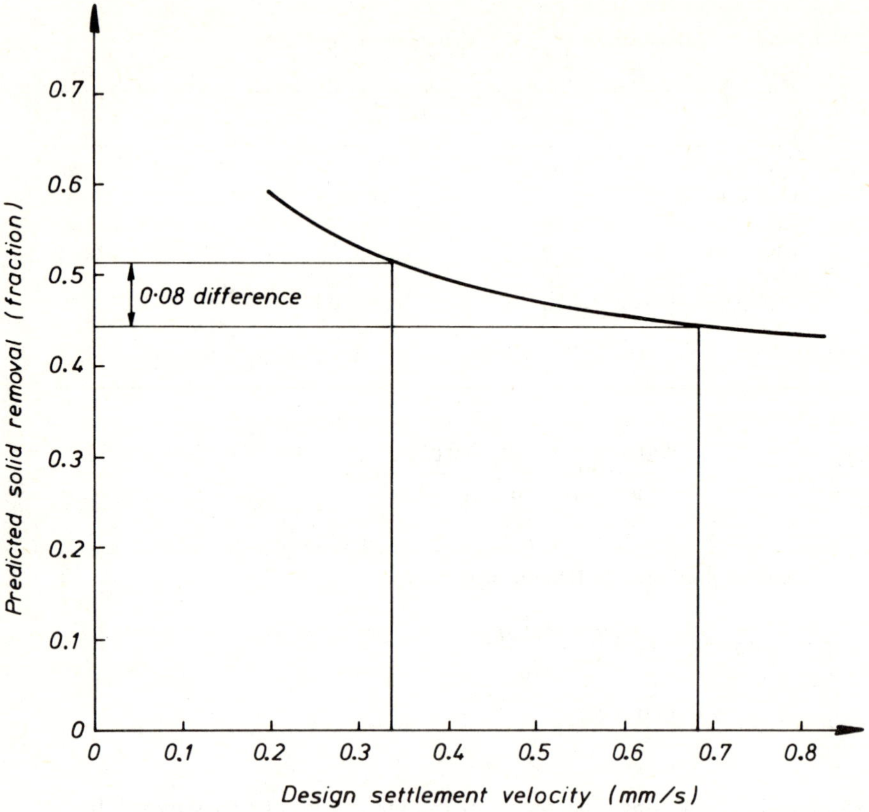

Fig. 3.4 Predicted suspended solids removal

Example 3.5

For the suspension tested in the previous section, determine the percentage removal of particles expected in an upward flow tank with an overflow rate of 26 m^3/m^2 day.

Calculation

Neglecting any liquid removed when desludging, the design settlement velocity V_0 is $(26 \times 10^3)/(24 \times 3600)$ mm/s, i.e. 0.3 mm/s.

From the batch settlement analysis given in Fig. 3.3, the value of $x_T = 1 - x_0$ would be 1 – 0.58 as a fraction or 42%. This compares with 53% for a horizontal flow tank.

□ □ □

3.3.2 Flocculent particles

If a particle or mass of particles form transient (weak) or permanent (strong) bonds with each other they are said to flocculate. Where particles flocculate (from Latin 'flocculus' – a tuft of wool) they increase their diameter with time of settlement. Describing the settlement characteristics of individual particles is therefore unrewarding and analysis of their behaviour is carried out by a batch settlement apparatus similar to that described in Section 3.2 and Fig. 3.2. However in the case of flocculent removals it is more common to present the results in the form of a grid showing percentage *removal* of particles at particular retention times and depths similar to that given in Fig. 3.5.

Example 3.6

Assuming the figures given in Fig. 3.5, calculate the theoretical percentage removal expected in a 2 m deep tank after 1 h retention and in a 2.5 m deep tank and 3.75 h retention.

Calculation

From Fig. 3.5 the removal at 2 m depth and 1 h is seen by inspection to be 34%. For 2.5 m and 3.75 h, arithmetic interpolation may be used, i.e.

$$\% \text{ removal} = \frac{1}{2} \quad \frac{(64+62)}{2} + \frac{(60+68)}{2} = 63.5\%$$

□ □ □

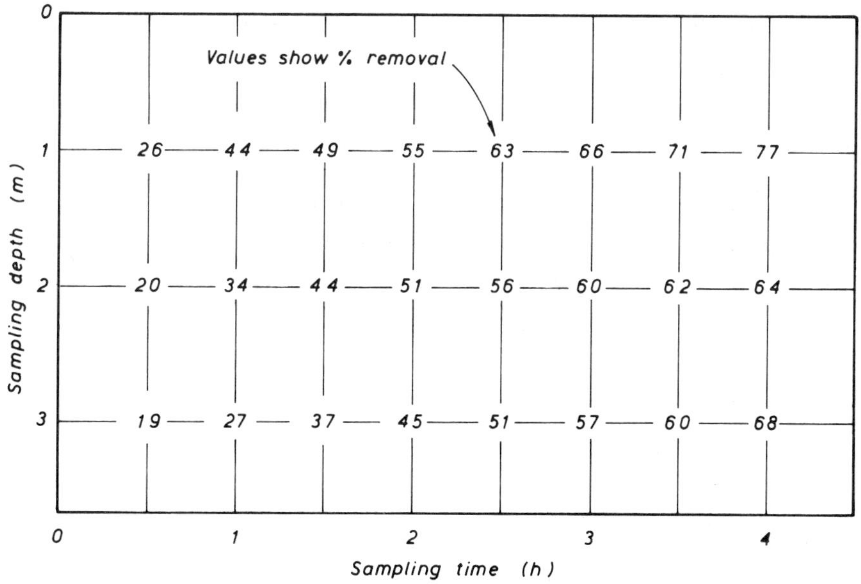

Fig. 3.5 Flocculent particles removal grid

3.4 Choice of tank shape

Although the foregoing calculations appear simple in theory, a question exists as to whether primary settlement can be described as the settlement of discrete particles (and thus primarily influenced by overflow rate and not by depth) or whether it is the settlement of flocculent particles (which implies that depth of tank or retention time are critical parameters). It is common therefore to design tanks such that both overflow rates and retention times are within conventional ranges. In addition, the overflow rate at the outlet weir has an influence on tank performance and this is also included in Table 3.4, which lists common ranges for these three criteria for a mid latitude country. In colder climates, the viscosity effect (see Equation (2.19)) decreases the required overflow rate and retention times while in equatorial climates a major problem is the prevention of the anaerobic conditions that would be caused by long retention times at elevated temperatures. The two commonest tank shapes are circular in plan and rectangular box shaped.

3.4.1 Radial flow tanks

Fig. 3.6 shows a circular settlement tank with the major dimensional relationships specified. These, together with the information from Table 3.4 can be used to determine tank dimensions.

Example 3.7

Calculate the approximate dimensions and number of tanks required to treat a maximum flow of raw sewage of 0.40 m^3/s.

Calculation

If we assume a peripheral (single) weir loaded at say 150 m^3/m day and the effective surface area of the tank (tank area less inlet area) to be loaded at 30 m^3/m^2day, then the maximum individual tank diameter can be found by equating (surface loading x area) to (weir loading x length). An allowance is made in the diameter d to allow for the inlet area, i.e.

$$\frac{30\pi(d - 0.15d)^2}{4} = \pi d\ 150$$

thus

$$d = \frac{150 \times 4}{30 \times 0.85^2} = 27.7 \text{ m}$$

which is equivalent to a flow of

$$\frac{30\pi 27.7^2\ 0.85^2 \text{ m}^3/\text{s}}{4 \times 8.64 \times 10^4} = 0.15 \text{ m}^3/\text{s}$$

Table 3.4 Typical operating criteria for settlement tanks in a temperate climate at maximum hourly flow rates, e.g. 3 DWF

Function	Retention time (h)	Loading rate ($m^3/m^2 d$)	Weir overflow rate (m^3/md)	Solids loading ($kg/m^2 d$)
Primary settlement				
Large scale: prior to activated sludge	$1-1\frac{1}{2}$	30–45	< 220	–
Large scale: prior to percolating filters	$1\frac{1}{2}-2$			
Small scale	> 15	–	–	–
Humus tanks				
large scale	$1\frac{1}{3}-1\frac{2}{3}$	35–40	< 220	–
small scale	> 4	–	–	–
Activated sludge				
large scale		45–60	< 250	< 180
small scale (extended aeration)		< 22	< 100	< 100

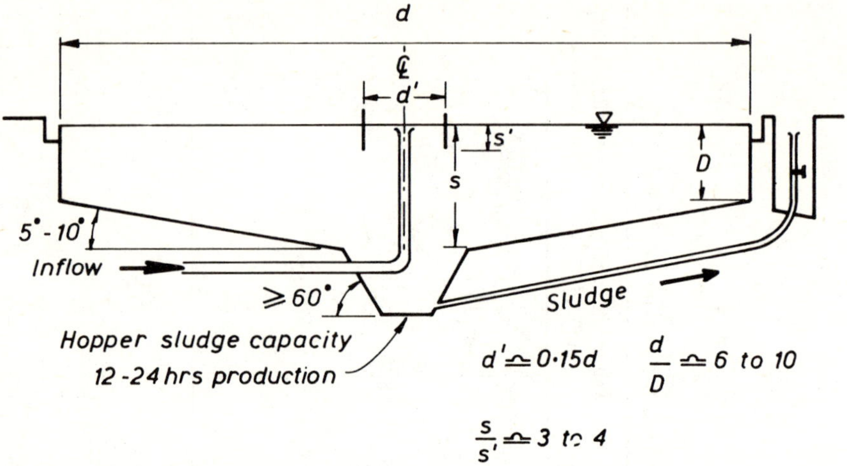

Fig. 3.6 Radial flow tank (by courtesy of Applied Science Publishers [5])

The number of tanks required is therefore 3. The actual diameter of the three tanks is given by

$$3\frac{\pi}{4}(0.85)^2 d_1^2\ 30 = 0.4 \times 8.64 \times 10^4$$

$$d_1 = 26.0 \text{ m}$$

To obtain a retention time of say 2 h at peak flow, the average depth, D, would be givin by

$$\frac{3\pi(0.85 \times 26.0)^2}{4}\frac{D}{0.4} = 2 \times 3600$$

i.e. $D = 2.5$ m. This would be satisfactory for a wall depth. If the bottom slope was 10° the actual retention time would be slightly greater than 2 h. The weir overflow rate (WOR) would be

$$\text{WOR} = \frac{0.4 \times 8.64 \times 10^4}{3\pi 26.0}$$

$$\text{WOR} = 140 \text{ m}^3/\text{m day}$$

Assuming an inlet concentration of solids of, say, 200 mg/l (0.2 kg/m^3) and 60% removal in the form of a sludge of 97% moisture content (30 kg/m^3), the size of a sludge hopper sufficient to hold 12 h sludge production would be

$$\text{volume of hopper} = \frac{0.4 \times 8.64 \times 10^4}{3 \times 2} \times \frac{0.20}{30} \times 0.6$$

$$= 23.0 \text{ m}^3$$

The volume of a cone whose apex angle is 60° is given by

$$\text{hopper volume} = \frac{\pi}{3}\, r_b^2\, h \tag{3.11}$$

where r_b is the hopper base (tank floor) radius and h is the hopper (or cone) depth. Since $r_b/h = \tan 30°$, then

$$\text{hopper volume} = 1.81 r_b^3$$

If the cone is truncated, with a bottom radius of 0.8 m then

$$1.81 r_b^3 - 1.81\,(0.8)^3 = 23$$

thus $\quad r_b = 2.4$ m

□ □ □

It should be noted that radial flow tanks usually have wall depths of 2.5 – 3.5 m for primary settlement and 3–4 m for secondary settlement. Tank diameters of up to 60 m have been used for primary settlement; however, they require an extra (inboard) weir and greater excavation. Bottom slopes between 5° and 10° are commonly used where scraper mechanisms are employed.

3.4.2 Rectangular tanks

Rectangular tanks are often used for primary settlement. They are not commonly chosen as activated sludge tanks since the efficient collection and return of such sludges is easier using a radial flow continuously scraped tank. For rectangular tanks common ratios and values of length, depth and width are given in Table 3.5.

Table 3.5 Dimensional ratios for rectangular tanks

$\frac{\text{Length}}{\text{Depth}}$	approximately 12/1 : length < 90 m
$\frac{\text{Length}}{\text{Width}}$	approximately 4/1
Depth using 3–3.5 m; maximum 5 m	
Retention time $1\frac{1}{2}$ to 2 h at maximum flow	
Minimum number of tanks = 2	

In practice many values of the ratios in Table 3.5 are adjusted to make better use of site conditions, this being easier in the case of rectangular tanks compared with radial flow tanks.

Example 3.8

Calculate approximate sizes for primary settlement tanks to treat a maximum flow of 0.5 m^3/s, using a design settlement velocity of 0.45 mm/s and a retention time (at maximum flow) of 1.75 h.

Calculation

$$\text{Surface loading rate} = 0.45 \times \frac{8.64 \times 10^4}{10^3}\ \text{m}^3/\text{m}^2\,\text{day} = 39\ \text{m}^3/\text{m}^2\,\text{day}$$

$$\text{The total area required} = \frac{0.5}{0.45 \times 10^{-3}}\ \text{m}^2 = 1110\ \text{m}^2$$

The depth is derived from the retention time

$$= \frac{\text{volume of tank}}{\text{inflow}}$$

$$= \frac{1110}{0.5} D \text{ or } 0.62\,D \quad \text{h}$$

Therefore

$$\text{depth} = \frac{1.75}{0.62} = 2.82\ \text{m; say 3 m allowing for sludge storage.}$$

Now $L/B = 4$; therefore

$$B = \frac{L}{4} \text{ and } \frac{nL^2}{4} = 1110\ \text{m}^2$$

A table of length, depth, etc. can be constructed for any value of n as in Table 3.6.

Table 3.6 Rectangular tank dimensions

	Length	Breadth	Weir overflow rate*	Forward velocity
n	$L = \sqrt{\left(\frac{4 \times 1110}{n}\right)}$ m	$B = \frac{L}{4}$ m	WOR = Q/nB m^3/m day	FV = $\frac{Q}{nBD}$ m/s
2	47.1	11.8	1830	7.1×10^{-3}
3	38.5	9.6	1500	5.8×10^{-3}
4	33.3	8.3	1301	5.0×10^{-3}
5	29.8	7.5	1152	4.4×10^{-3}
6	27.2	6.8	1058	4.1×10^{-3}

* assuming a single sided weir across the end of the tank.

□ □ □

The decision as to the number of tanks depends on the site conditions; however, reducing the number of tanks generally reduces the duplication of mechanical scraping equipment while increasing the weir overflow rate. If four tanks 33.3 m x 8.3 m were chosen the WOR would be 1300 m^3/m day. Although some argument is common regarding the necessity of reducing this value to 200 m^3/m day, the former figure is clearly too high. Installing two double sided inboard weirs supported across the tank would, together with the end weir, reduce the WOR to 260 m^3/m day.

(a) Rectangular tank design – Clements method
A full description of this approach to design for both rectangular and circular settlement tanks is given elsewhere [1]. However, this process effectively seeks relationships between the various tank dimensions. The steps involve:

(i) rational choice of overflow rate or design settlement velocity V_0, which leads to a choice of length/depth ratio in terms of the mean forward velocity V since $V_0/V = D/L$ from before.

(ii) a suggestion by Clements that if $L^2/BD > 20$ then the magnitude of the maximum forward velocity at the bottom of the tank will be less than $8V$. Thus if V is set at 1/8 the scouring velocity of the sludge, there ought to be minimal scouring or resuspension of deposited solids.

(iii) allowance being made for poor velocity distribution in plan by means of a 'time ratio factor' which increases the surface area of the tanks, i.e.

$$\text{area} = \frac{\text{maximum flow}}{T_R V_0} \qquad (3.12)$$

where T_R has values between 0 (infinitely poor tank) and 1 (ideal tank). Ratios of $L/B > 2$ tend to improve the value of T_R.

(iv) from a consideration of the construction costs, ratios of length/total width of all tanks should be between 1/2 and 1.

Example 3.9
Using the Clements' approach, design horizontal flow settling tanks for primary sewage (scouring velocity 32 mm/s). Maximum flow is 0.1 m^3/s; V_0 is to be 0.6 mm/s and $T_R = 0.6$

Calculation
Let V be 1/8 scouring velocity, i.e. 32/8 or 4 mm/s. V_0 is 0.6 mm/s; thus

$$\frac{V}{V_0} = \frac{4}{0.6}$$

or thus

$$\frac{L}{D} = 6.7$$

Note that since this ratio is now set, any choice of tanks which increases the length will necessarily increase excavation costs.

The surface area

$$A = \frac{Q_{max}}{T_R V_0} = \frac{0.1}{0.6 \times 0.6 \times 10^{-3}} = 278 \text{ m}^2$$

If $B_0 = nB$, i.e. B_0 is the total breadth of n individual tanks of breadth B m; then from constructional considerations L/B_0 should range from 1/2 to 1.

The surface area is LB_0; thus L should range from $\sqrt{(278/2)}$ m to $\sqrt{278}$ m, i.e. 11.8 m to 16.7 m. B_0 is therefore between 278/16.7 m and 278/11.8 m, or 16.6 m and 23.6 m.

Since

$$\frac{L^2}{BD} > 20 \text{ then } \frac{nL^2}{B_0 D} > 20$$

and since L/D is set at 6.7

$$\frac{nL}{B_0} > 3.0$$

□ □ □

The results of such calculations are shown in Table 3.7 for various choices of L. If the values of n and B_0 corresponding to the particular choice of L are plotted as in Fig. 3.7, three points are possible where the number of tanks is an integer value, i.e.

$$n = 3, B_0 = 16.7 \text{ m}; \; L = 16.6 \text{ m and } L/B = 3.0$$

$$n = 4, B_0 = 19.4 \text{ m}; \; L = 14.3 \text{ m and } L/B = 3.0$$

$$n = 5, B_0 = 21.6 \text{ m}; \; L = 12.9 \text{ m and } L/B = 3.0$$

Table 3.7 Values of tank dimensions for various values of L

L(m)	12.0	13.0	14.0	15.0	16.0	17.0
B_0	23.2	21.3	19.8	18.5	17.4	16.3
D	1.80	1.94	2.09	2.24	2.39	2.54
n	5.8	4.9	4.2	3.7	3.3	2.9
n (rounded)	6	5	5	4	4	3
B (based on rounded n)	3.9	4.3	4.0	4.6	4.4	5.4
$\frac{L}{B}$	3.1	3.0	3.5	3.3	3.6	3.1

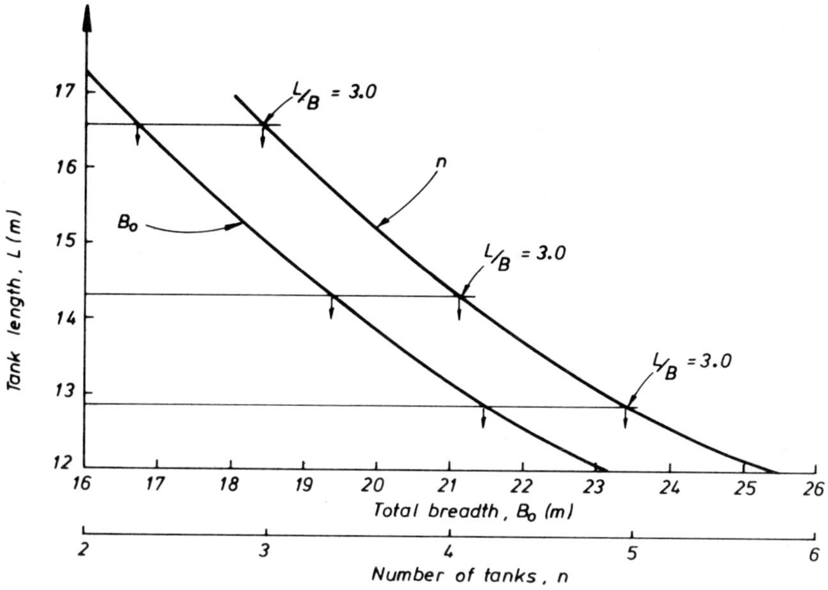

Fig. 3.7 B_0 and n as a function of L

The final choice could be made on site conditions; however, since increasing the tanks also increases the equipment cost, there is a tendency to choose three tanks, 16.6 m long and 5.5 m wide. The values of L/B have been included since values significantly less than 3.0 tend to reduce the values of T_r and thus produce a penalty by increasing the value of LB_0.

□ □ □

3.5 Settlement efficiency – primary tanks

Tebbutt [2] has proposed an equation to calculate the fractional removal efficiency E in a primary settlement tank with overflow rate Q/A and with influent suspended solids concentration S_i, i.e.

$$E = 0.955 \exp - [265/S_i + 0.0021\ Q/A] \tag{3.13}$$

The equation implies that the efficiency is more likely to be influenced by changes in S_i than V_0 for commonly used values of both parameters.

Example 3.10
Using Equation (3.13) determine the expected removal efficiency (*a*) for an overflow rate of 25 m^3/m^2 day and S_i values of 300 and 600 mg/l, and (*b*) for an S_i value of 400 mg/l, for overflow rates of 25 and 100 m^3/m^2 day.

Calculation

(*a*)i for $Q/A = 25\ \mathrm{m^3/m^2}$ day and $S_i = 300$ mg/l

$$E = 0.955 \exp - \left[\frac{265}{300} + 0.0021 \times 25\right] = 0.37$$

ii for $Q/A = 25\ \mathrm{m^3/m^2}$ day and $S_i = 600$ mg/l

$$E = 0.955 \exp - \left[\frac{265}{600} + 0.0021 \times 25\right] = 0.58$$

i.e. a change of 21%.

(*b*) For influent solids of 400 mg/l the difference in E for flows of 25 and 100 $\mathrm{m^3/m^2}$day is

$$\Delta E = 0.955 \exp - \left[\frac{265}{400} + 0.0021 \times 25\right] - 0.955$$

$$\exp - \left[\frac{265}{400} + 0.0021 \times 100\right]$$

$$\Delta E = 0.47 - 0.40 = 0.07 \text{ or } 7\%$$

□□□

3.6 Activated sludge settlement

The activated sludge process is described more fully in Chapter 4. Briefly, the sludge is a largely organic material which is flocculent and free floating in an aeration tank where it is contacted with sewage (see Section 4.1.1 and Fig. 4.1). The sewage is mixed with returned settled sludge prior to aeration for several hours. The mixed liquor suspended solids (MLSS), as the mixture of tank solids is called, enters the final settlement tank as part of a continuous process. As Fig. 4.1 shows, there is an underflow of thickened sludge which is returned to be mixed with fresh sewage.

In the activated sludge settlement tanks, two processes take place:

(*a*) the clarification of the liquor (effluent) which goes over the outlet weir
(*b*) the thickening of the sludge prior to return to the aeration tank.

3.6.1 Clarification

If a sample of activated sludge is shaken in a wide, deep tube (say 1 m x 0.01 m) it will settle with a distinct interface between sludge and liquor (effluent) and the plot of interface height versus time often forms a curve of the type shown in Fig. 3.8. The percentage of the original volume of the column which the

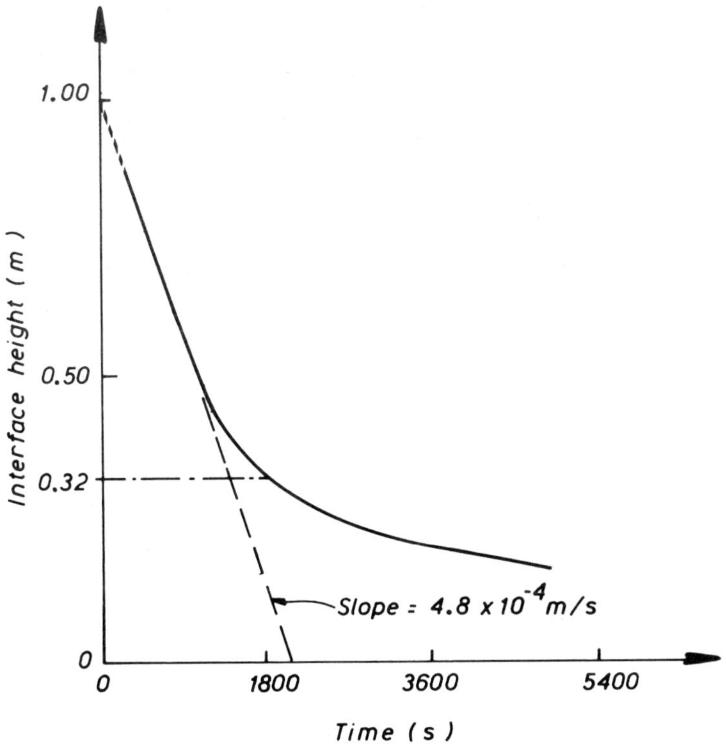

Fig. 3.8 Clarification curve

sludge occupies after 30 min settlement divided by the initial sludge concentration (MLSS) % gives the sludge volume index SVI, i.e.

$$\mathrm{SVI} = \frac{\text{Volume occupied by sludge as \% of original volume}}{\text{MLSS \%}} \tag{3.14}$$

This is used to give a qualitative estimate of the sludge settleability. The inverse of the SVI value multiplied by 100 gives the sludge density index, SDI. Values of 150–200 for the SVI are poor and values of $\leqslant 50$ are good. From Fig. 3.8 it can be seen that the rate of production of clarified effluent Q_e is given by the product of the initial clarification velocity V_I and the surface area A.

In a real tank, i.e. in Fig. 3.9, there is an inlet flow $(Q_e + q)$ at concentration C_0 and two outflows (i)Q_e with what is hopefully a negligible concentration of solids C_e, and (ii) an underflow q of solids concentration C_u. The example below illustrates how the overflow rate concept may be used to determine a tank surface area.

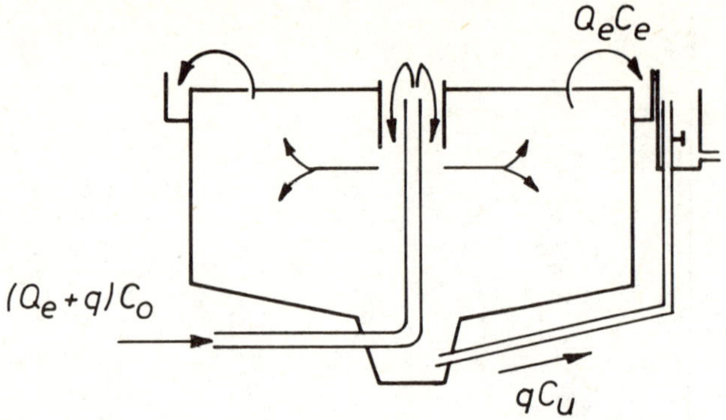

Fig. 3.9 Activated sludge settlement tank solids balance

Example 3.11
Assuming the sludge settlement curve given in Fig. 3.8 is representative of a mixed liquor of MLSS (C_0) = 3500 mg/l which is fed into an activated sludge settlement tank at 0.02 m^3/s, calculate the tank surface area required *for clarification* if thickened sludge is withdrawn at a concentration C_u = 13 000 mg/l. Also calculate the SVI of the sludge.

Calculation
From a mass balance in Fig. 3.9

$$(Q_e + q)C_0 = Q_eC_e + qC_u \qquad (3.15)$$

If $C_e \cong 0$ (it usually is < 30 mg/l) then

$$\frac{Q_e}{q} = \frac{C_u - C_0}{C_0} \qquad (3.16)$$

i.e.

$$\frac{Q_e}{q} = \frac{13\,000 - 3500}{3500} = 2.71$$

Therefore

$$Q_e = \frac{2.71}{3.71} \times 0.02 \text{ m}^3/\text{s} = 0.015 \text{ m}^3/\text{s}$$

Then

$$V_IA = Q_e$$

V_I from Fig. 3.8 is 4.8×10^{-4} m/s, therefore

$$\text{the area required} = \frac{0.015}{4.8 \times 10^{-4}} = 31 \text{ m}^2$$

Also, from Fig. 3.8, the height of the sludge column after 30 min is 320 mm. The percentage settled volume after 30 min

$$= \frac{A \times 320}{A \times 1000} \times 100\% = 32\%$$

The initial concentration C_0 = 3500 mg/l which can be expressed as approximately 3500 p.p.m. (parts per million) since the density of the inflow is almost 1000 kg/m^3, i.e.

$$C_0 = 0.35\%; \ (10\,000 \text{ p.p.m.} = 1\%)$$

Thus

$$\text{SVI} = \frac{32}{0.35} = 91$$

i.e. the sludge is of medium settleability.

□ □ □

3.6.2 Activated sludge thickening

One of the most powerful theories relating to the thickening of activated sludges is that of flux thickening[3, 4]. It is based on the assumption that as a sludge thickens, i.e. changes its concentration C, the average velocity of that mass of sludge V is related to C by the expression

$$V = V^1 e^{-kC} \qquad (3.17)$$

where V^1 and k are constants. In a batch settlement tank of area A the rate of transfer of solids downwards through a section 'j' is equal to the product of V_j and C_j.

For a settlement tank with a continuous input, weir overflow and sludge return, the underflow q draws down solids through the section at a rate which is given by the product of q/A and C_j. Thus the overall transfer of solids through the section j is given by

$$\psi_j = V_j C_j + \frac{qC}{A} j \qquad (3.18)$$

q/A is usually designated 'U' being the underflow resultant velocity. If C_e is negligible, then for a steady state, the input solids flux

$$\frac{(Q_e + q)C_0}{A}$$

must equal ψ_j. If the input flux is greater than ψ_j, material will overflow, i.e. C_e is not negligible. The effect of varying q/A is best seen by an example.

Example 3.12
If C is given as kg/m^3 and V is m/h

$$V = 6e^{-0.4C}$$

is a realistic description of the mass settling velocity of a particular activated sludge. Plot the V versus C curve and hence construct the flux settlement curve for underflow values (q/A) of 0.3 m/h and 0.8 m/h.

Calculation
For $C = 0$ kg/m^3, $V = 6.0$ m/h. For $C = 1$ kg/m^3 (1000 mg/l), $V = 4.0$ m/h.
Similarly, for various values of C, Fig. 3.10 can be constructed. Pairs of values for V and C can be calculated and the product $V \times C$ can be estimated (Table 3.8) and plotted (Fig. 3.11). If the value of q/A or U is now chosen, a cumulative flux equal to $VC + UC$ can also be determined and plotted as in Fig. 3.12

Table 3.8 Settlement flux values VC and $VC + UC$

C (kg/m^3)	VC (kg/m^{2}h)	$VC + UC$ $U = 0.3$ m/h	$U = 0.8$ m/h
0.0	0.0	0.0	0.0
2.0	5.4	6.0	7.0
4.0	4.8	6.1	8.1
6.0	3.3	5.1	8.1
8.0	2.0	4.4	8.4
10.0	1.1	4.1	9.1
12.0	0.6	4.2	10.2
14.0	0.3	4.5	11.5
16.0	0.16	5.0	13.0

□□□

Considering curve '*a*' in Fig. 3.12, i.e. $U = 0.3$ m/s, the compound flux curve shows a maximum at approximately $C = 2.5$ kg/m^3 and a minimum at $C \cong$ 10.4 kg/m^3. If the MLSS enters the tank at concentration $C_0 = 4$ kg/m^3, then as the sludge thickens or increases in concentration, the ability of the tank to pass solids downwards diminishes until a value of 10.4 kg/m^3 is reached and subsequently increases. The minimum compound flux ψ_L (i.e. 4.1 kg/m^{2}h) imposes a limit to the solids handling capacity of the thickener.

If the solids concentration in the effluent stream, C_e, is negligible, then the inlet solids flux, ψ_I, must equal the return sludge solids flux, $(q/A)C_u$ unless there is storage or hold up of solids in the settlement tank. Where no hold up of

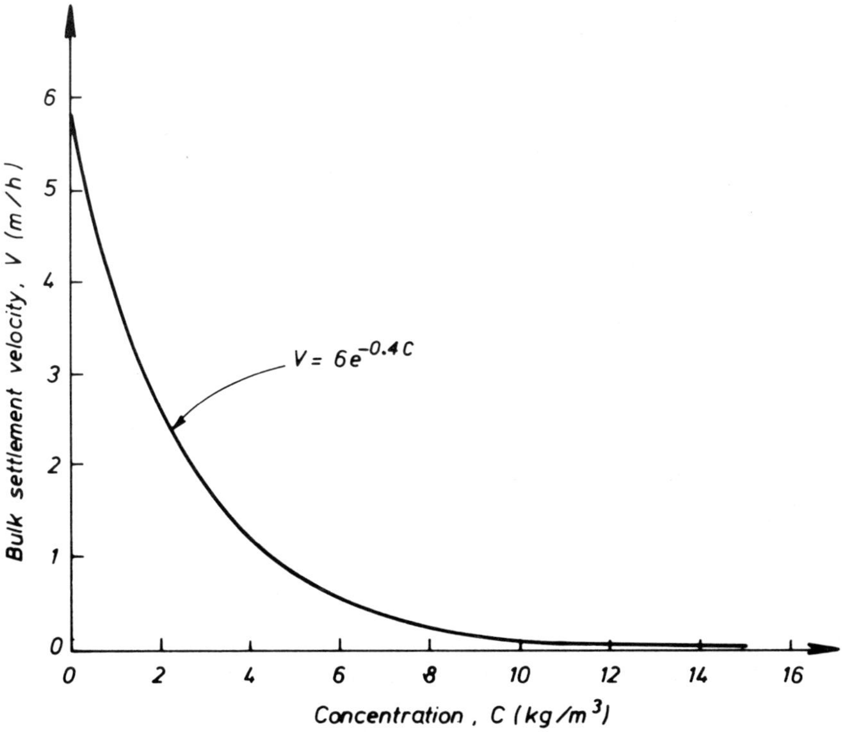

Fig. 3.10 V versus C curve

solids occurs, i.e.

$$\psi_I = \frac{(Q_e + q)C_0}{A} = \frac{q}{A} C_u = UC_u \tag{3.19}$$

this relationship can be used to estimate C_u.

Example 3.13

For curve '*a*' in Fig. 3.12, calculate the value of C_u for inlet flux loadings of 4.1, 3.0 and 6 kg/m^{2}h.

Calculation

For $\psi_I = \psi_L$ = 4.1 kg/m^{2}h, this is equal to the solids handling capacity of the thickener and there should be no solids hold up; thus

$$C_u = \frac{\psi_I}{U} = \frac{4.1}{0.3} \text{ or } 13.7 \text{ kg/m}^3$$

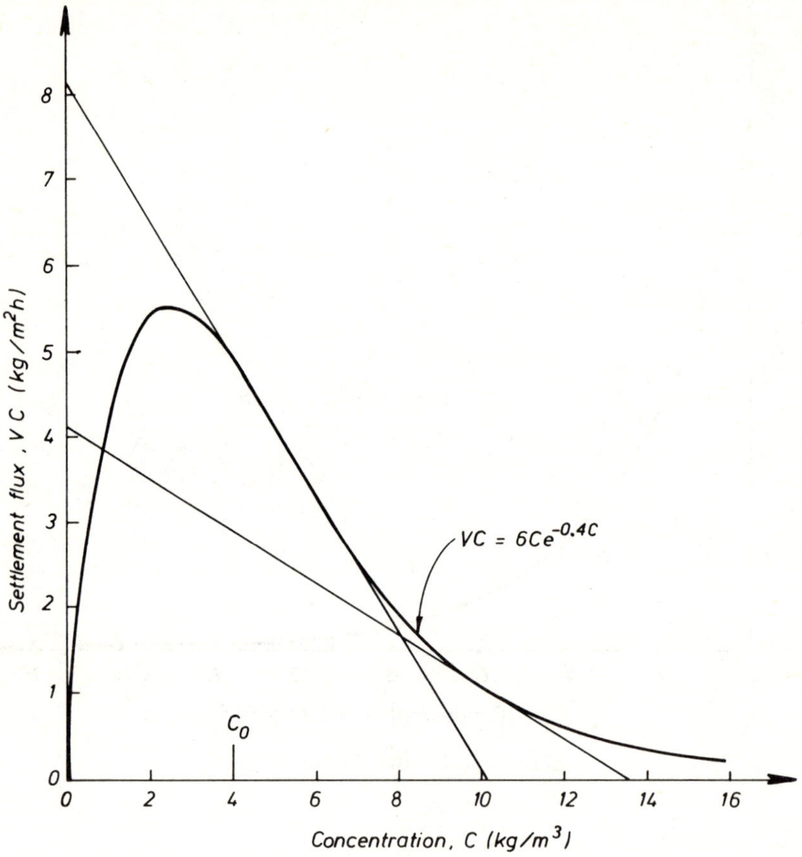

Fig. 3.11 $V \times C$ versus C curve

as shown in Fig. 3.12. Similarly for $\psi_I = 3.0$ kg/m^{2}h

$$C_u = \frac{3.0}{0.3} \text{ or } 10.0 \text{ kg/m}^3$$

For $\psi_I = 6.0$ kg/m^{2}h, the thickener can only handle 4.1 kg/m^{2}h (at this value of U.) Thus the outlet flux $(q/A)U$ must equal 4.1 kg/m^2 and C_u = 13.7 kg/m^3 as before.

□□□

In this latter case the thickener is being overloaded at a rate of (6.0 – 4.1) kg/m^{2}h or 1.9 kg/m^{2}h. If these conditions persist eventually this amount of solids will be lost in the effluent stream. The condition may be relieved by

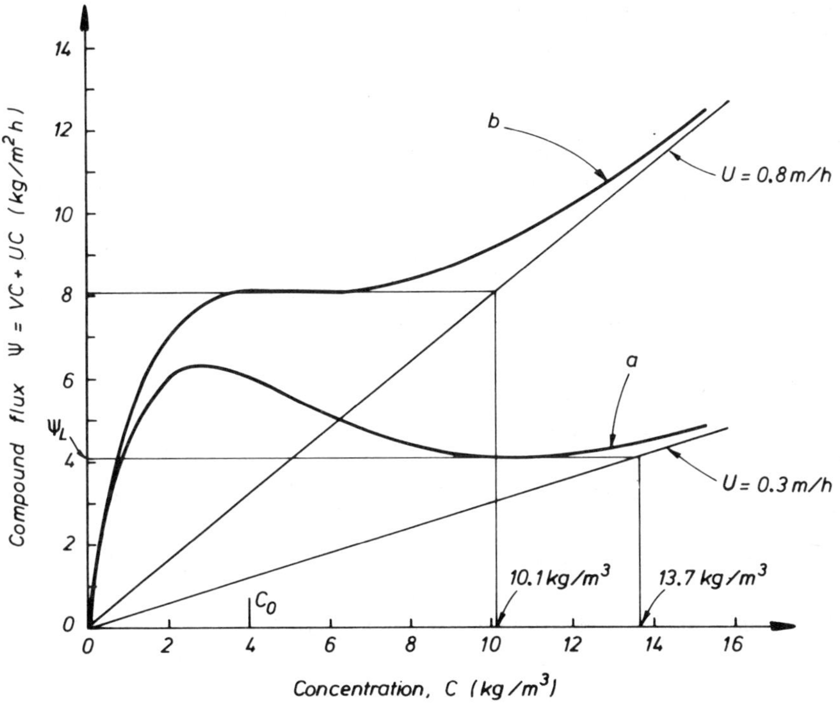

Fig. 3.12 Compound flux curve

increasing U. If U is increased to 0.6 m/h, the value of ψ_L increases to 6.7 kg/m^{2}h but now occurs at a concentration of 7.5 kg/m^3. The value of C_u for an inlet flux of 6.0 kg/m^{2}h then is 6.0/0.6 or 10 kg/m^3. Now considering curve '*b*' in Fig. 3.12, as U attains the value of approximately 0.8 m/h there is merely a point of inflection in the combined flux curve, i.e. there is no minimum flux level ψ_L. As before $C_u = \psi_I/U$; however, the limiting criteria for the tank is now its role as a clarifier not as a thickener.

Example 3.14
Assuming $V = 6 \exp(-0.4C)$ describe the thickening behaviour of the sludge, show by both general formulae and by calculation, that clarification is critical when $U = 0.8$ m/h (approximately).

Calculation
If solids are not to be lost in the effluent stream

$$\psi_I \leqslant \psi_L \tag{3.20}$$

Now

$$\psi_I = \frac{C_0(Q_e + q)}{A} \tag{3.21}$$

From Fig. 3.12, ψ_L for $U = 0.8$ m/h has become in fact the value of ψ at $C = C_0$ (i.e. 8.1 kg/m^{2}h)

$$\psi_L = V_I C_0 + \frac{q}{A} C_0 \tag{3.22}$$

where V_I is the initial settlement velocity of the solids at inlet concentration C_0. By comparison of the above two expressions

$$V_I \geqslant \frac{Q_e}{A} \tag{3.23}$$

which is the criterion for clarification as shown in Section 3.6.1. When $U = 0.8$ m/h, an inlet flux of 8.1 kg/m^{2}h gives an underflow concentration C_u of

$$C_u = \frac{8.1}{0.8} = 10.1 \text{ kg/m}^3$$

Therefore, for an inlet MLSS concentration of 4 kg/m^3

$$\frac{Q_e}{q} = \frac{C_u - C_0}{C_0} = \frac{10.1 - 4}{4} \text{ or } 1.5$$

and $\frac{Q_e}{A} = 1.5 \times 0.8$ or 1.22 m/h.

Thus, at a value of $U = 0.8$ m/h the effluent is being withdrawn at a rate equivalent to an upward flow velocity of 1.22 m/h for $C_0 = 4$ kg/m^3.

However, the initial settling velocity of the sludge bulk is given by

$$V_I = 6 \exp(-0.4\, C_0)$$

i.e.

1.21 m/h

i.e. the clarification rate criteria that the overflow velocity be less than the initial sludge settling velocity has been transgressed (the velocities are actually equal for $U = 0.79$ m/h).

□ □ □

Although the production of compound flux curves of the type shown in Fig. 3.12 is made easy by use of computers, an easier method of determining changes in C_u owing to variations in U is possible using Fig. 3.11. If a line is drawn under the flux curve with gradient $-U$, then the intercept on the ordinate

(ψ value) is equal to the underflow concentration $C_u \times U$ as before. It is also equal to

$$\psi = V_j C_j + \frac{q}{A} C_j \text{ or } V_j C_j + UC_j \tag{3.24}$$

where C_j is the concentration at that section. When the line touches the curve, the ordinate value is the maximum flux loading rate ψ_L and the abcissa intercept is the maximum C_u value. When U is increased to approximately 0.8 m/h, the line cuts the curve at a C value of C_0, i.e.

$$\psi_L = V_0 C_0 + UC_0 \tag{3.25}$$

which is again the condition at which clarification criteria override thickening criteria. The method only involves a straight edge, to determine different values of ψ_L for each value of U.

Example 3.15
Using Fig. 3.11, construct the line with slope $-U$ for $U = 0.3$ m/h and find C_u, ψ_L and the concentration at which ψ_L occurs. Also determine the value of C_u for an input flux of 6 kg/m^{2}h and a U value of 0.6 m/h.

Calculation
If a line is drawn joining 3.0 kg/m^{2}h on the ordinate and 10.0 kg/m^3 on the abcissa, i.e. a slope of -0.3 m/h, and a rule moved parallel to this line such that it becomes tangential to the curve $VC = 6Ce^{-0.4C}$, then the abcissa value of 13.7 kg/m^3 is C_u. The concentration at which ψ_L occurs is the value of C at the tangent point, i.e. 10.3 kg/m^3 approximately. The ordinate value ψ_L is 4.1 kg/m^{2}h as before.

If an inlet flux of 6 kg/m^{2}h is to be treated, then raising the angle of the rule to give a U value of 0.6 m/h gives a line tangential to the curve at $C = 7.8$ kg/m^3 and a ψ_L value of 6.8 kg/m^{2}h and a maximum C_u value of 11.3 kg/m^3. The actual C_u value for an input flux of 6 kg/m^{2}h is as before:

$$\frac{6.0}{0.6} \text{ or } 10 \text{ kg/m}^3$$

□ □ □

The area of the thickeners required for a particular process can be calculated if a curve such as that shown in Fig. 3.11 can be constructed.

Example 3.16
A *works* inflow of 0.1 m^3/s produces an MLSS of concentration 4000 mg/l from an activated sludge unit. The MLSS is to be thickened to 12 000 mg/l in a

thickener. Calculate the area required and the loading rate if the sludge settling characteristics are described by the curve in Fig. 3.10. The settled sludge is to be returned to the inlet of the aeration tank.

Calculation
The value of Q_e is 0.1 m^3/s or 360 m^3/h, $C_u = 12$ kg/m^3 and $C_0 = 4$ kg/m^3. Therefore

$$\frac{q}{Q_e} = \frac{C_0}{C_u - C_0} = 0.5$$

Thus, the return sludge flow is given by

$$q = 0.5 \times 360 \text{ or } 180 \text{ m}^3/\text{h}$$

The actual inlet load is given by

$$(Q_e + q)C_0 = 2160 \text{ kg/h}$$

In Fig. 3.11, aligning a rule at $C = 12$ kg/m^3 to cut the curve tangentially, gives an ordinate value for ψ_L of 5.8 kg/m^{2}h and a slope $(-U)$ value of 5.8/12 or 0.48 m/h.

Using this loading rate and letting A be the required area

$$\psi_I \leqslant \psi_L$$

$2160/A \leqslant 5.8$, i.e. $A > 372$ m^2.

Using the underflow velocity as a check

$$\frac{180}{A} = 0.48; \; A = 375 \text{ m}^2$$

The overall loading rate using 372 m^2 and 360 m^3/h is (360/372) x 24 m^3/m^{2}day or 23.2 m^3/m^{2}day. The loading rate is commonly expressed in terms only of the works inflow.

□ □ □

Questions

Question 3.1

The results of a batch settlement test on a discrete particle suspension are given on the next page.

Construct the cumulative batch settlement curve, and, using Equation (3.8), calculate the removal expected in a horizontal flow tank of surface area 100 m^2 for flows of 0.08 m^3/s and 0.02 m^3/s.

Sampling depth (m)	Sampling time (h)	Suspended solids remaining in sample (%)
1	1	61
	3	43
	6	30
2	1	64
	3	56
	8	36
3	1	66
	2	63
	6	51

Question 3.2

Calculate the settlement velocity of spherical sand grains 0.5 mm and 2 mm diameter having relative density (specific gravity) of 2.65 settling in water of kinematic viscosity 1.3 x 10^{-6} m^2/s.

Question 3.3

The bulk settlement velocity of an activated sludge varies with concentration as in the figure below. Construct the mass flux versus concentration curve and find the limiting flux ψ_L for a sludge underflow concentration of 10 000 mg/l. Also calculate the area required for thickening 0.1 m^3/s of similar activated sludge at 3000 mg/l MLSS.

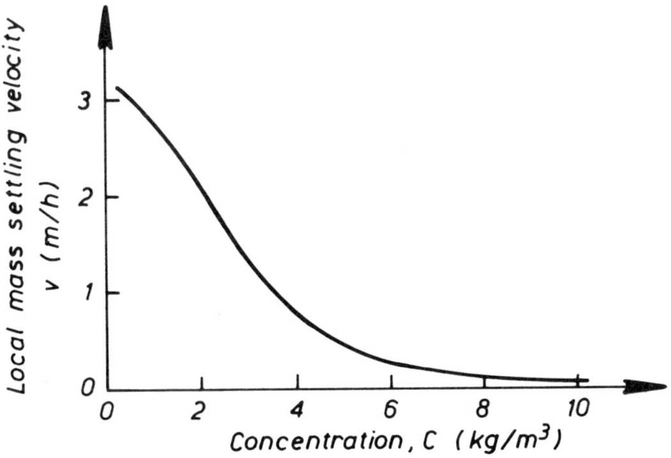

Question 3.4

The bulk settlement velocity (V, m/h) of an activated sludge is related to the concentration (C, kg/m^3) thus:

$$V = 6 \exp(-0.4C)$$

Construct the combined flux settlement curve for an underflow rate ($u = q/A$) of 0.3 m/h. For $u = 0.3$ m/h, find the value of the underflow concentration (C_u) for each of the three input fluxes (*a*) 2.5 kg/m^{2}h (*b*) 3.0 kg/m^{2}h (*c*) 5.0 kg/m^{2}h.

Question 3.5

A sewage treatment works is to have primary settlement using rectangular (horizontal flow) tanks. If the maximum flow to be treated is 0.5 m^3/s and the scouring velocity of the sludge is 32 mm/s, *design* settlement tanks according to the method proposed by M. S. Clements. Use a time ratio of 0.8, a design settlement velocity of 0.5 mm/s and a length of 25 m.

References

1 Clements, M. S. (1968), *Surveyor and Local Govt. Technol.*, Oct. 19th, 41–45.
2 Tebbutt, T. H. Y. and Christoulas, D. G. (1975), *Wat. Res.*, **9**, 347–56
3 Dick, R. L. and Ewing, B. B. (1967), *J. San. Engng. Divn. A.S.C.E.*, SA 4, **96**, 9–29.
4 White, M. J. D. (1975), *Settling of Activated Sludge*. Water Research Centre Technical Report No. TR 11, WRC, Medmenham, UK.

Further reading

Barnes, D. and Wilson, F. (1978), *Chemistry and Unit Operations in Sewage Treatment*, Applied Science Publications, London, Chapter 5.

4

Secondary treatment processes

Nomenclature

Symbol	*Definition and units*	*Dimensions*
μ	Specific growth rate, h^{-1} or day^{-1}	T^{-1}
$\hat{\mu}$	Maximum value of specific growth rate, h^{-1} or day^{-1}	T^{-1}
K_s	Constant, equal to S_e when $\mu = \hat{\mu}/2$, mg/l of COD or BOD as stated	ML^{-3}
S_e	Substrate (food) source, mg/l of COD or BOD as stated	ML^{-3}
X	Biomass concentration or bacterial concentration, mg/l	ML^{-3}
t	General symbol for time of growth or retention time, h, day etc.	T
X_0	Initial value of X at $t = 0$ or mixed liquor suspended solids concentration, mg/l	ML^{-3}
t_g	Generation or doubling time for bacteria, h or day	T
S	Substrate concentration (general), mg/l	ML^{-3}
Y	Yield of biomass per unit of substrate removed, dimensionless	–
L_r	BOD_5 removed in process, mg/l	ML^{-3}
F	Inflow to tank from primary stage, l/s, m^3/s, m^3/day etc. as stated	L^3T^{-1}
q	Returned sludge flow, l/s m^3/s, m^3/day etc. as stated	L^3T^{-1}
X_u	Returned sludge concentration, mg/l	ML^{-3}
L_e	BOD in effluent stream, mg/l	ML^{-3}
L_i	BOD in influent stream, mg/l	ML^{-3}
J	Exponential constant, implied units, 1/h mg	$L^3T^{-1}M^{-1}$
K_d	Decay coefficient for total sludge solids, day^{-1}	T^{-1}
X_v	Volatile biomass concentration, mg/l	ML^{-3}

X_{v0}	Mixed liquor volatile suspended solids	ML^{-3}
K'_d	Decay constant for volatile sludge solids, day^{-1}	T^{-1}
F_w	Waste sludge flow, l/s, m^3/day as stated	L^3T^{-1}
V	Volume of aeration tank, filter media, oxidation pond, l or m^3	L^3
θ_c	Sludge age = (sludge mass, kg/sludge wasted, kg/day)	T
SLR	Sludge loading ratio; BOD input, kg/day/sludge mass, kg	T^{-1}
SUR	Sludge utilization ratio; BOD removed, kg/day/volatile sludge mass, kg	T^{-1}
Y	Biomass yield coefficient, dimensionless	–
X_{vi}	Biomass volatile solids in input stream, mg/l	ML^{-3}
X_{ve}	Biomass volatile solids in effluent stream, mg/l	ML^{-3}
X_{vw}	Biomass volatile solids in waste stream, mg/l	ML^{-3}
W_V	Waste sludge, kg/day	MT^{-1}
M_B	BOD removed, kg/day	MT^{-1}
M_v	Total sludge solids, kg	M
K	Constant = 1 day^{-1}	T^{-1}
T	Temperature	°C
C_s	Saturation concentration of oxygen, mg/l	ML^{-3}
m	Mass of oxygen transferred, mg, g or kg as stated	M
K_La	Proportionality constant, includes bubble area in numerator and tank volume in denominator, h^{-1} or day^{-1}	T^{-1}
C	Oxygen concentration, mg/l	ML^{-3}
D	Dissolved oxygen deficit ($D = C_s - C$), mg/l	ML^{-3}
D_0	Dissolved oxygen deficit when $t = 0$ mg/l	ML^{-3}
α	Constant to correct K_La for wastewater impurities, dimensionless	–
β	Constant to correct C_s for wastewater impurities, dimensionless	–
P	Experimental pressure, kN/m^2	$ML^{-1}T^{-2}$
p_w	Pressure exerted in gas phase by water molecules, water saturation pressure, kN/m^2	$ML^{-1}T^{-2}$

Z	Altitude above mean sea level, m	L
OC	Oxygenation capacity of aerator, kg/h, mg/h etc. as stated	MT^{-1}
E	BOD removal efficiency, fraction or % as stated	–
W	BOD load to filter, kg/day	MT^{-1}
f	Recirculation factor $f = \left(1 + \frac{R}{F}\right) \Big/ \left(1 + 0.1\frac{R}{F}\right)^2$	–
R	Recirculated flow to filter, units as F	L^3T^{-1}
A_s	Specific surface area of filter media. Area per (gross) unit volume, m^2/m^3	L^{-1}
A	Rotating disc area, both sides of disc counted, or surface area of pond, m^2	L^2
Q	Corrected flow rate to discs, m^3/day	L^3T^{-1}
L_D	BOD input load, mg/day	MT^{-1}
T_m	Mean water temperature in pond in coldest month, °C	°C
k	Bacterial decay constant in pond, day^{-1}	T^{-1}
N_t	Number of bacteria/100 ml in effluent stream or at time t	–
N_0	Number of bacteria/100 ml in effluent stream or at time $t = 0$	–
n	Number of ponds in series	–

4.1 Introduction

In many wastewater treatment plants the preliminary and primary stages can efficiently remove 30–40% of the influent BOD and 60–75% of influent SS. Those fractions remaining are soluble, colloidal or sufficiently small not to settle easily, and they consist of a wide range of organic and inorganic materials. The usual way in which the polluting load is further reduced is to encourage micro-organisms to oxidize the organic material in a similar manner to that in the natural processes occuring in rivers and streams but at an increased rate. The treatment plant is therefore designed to provide a suitable environment in which the aerobic organisms can grow rapidly. A brief description of the organisms commonly involved in sewage treatment is given in Appendix 3, however, in many situations it is sufficient to limit the theoretical discussion to bacterial organisms. The conditions under which they thrive or fail are also briefly tabulated in Appendix 3. Three main types of secondary units are found:

4.1.1 Activated sludge

If sewage is aerated, a dark brown, earthy smelling floc of free floating 'activated sludge' organisms, about 70–80% volatile (organic) matter, is built up. The organisms oxidize the organic molecules using naturally produced enzymes which they synthesize from the organic substrate (food source.) In practice the process is operated as a continuous scheme with the sludge being aerated at high sludge concentrations (mixed liquor suspended solids or MLSS). The sludge is separated after aeration and returned to be remixed with settled or unsettled sewage as shown in Fig. 4.1. As excess sludge builds up in the process some must be wasted. Extended aeration, contact aeration, pure oxygen, deep shaft and oxidation ditches (not ponds) are variations of the activated sludge process.

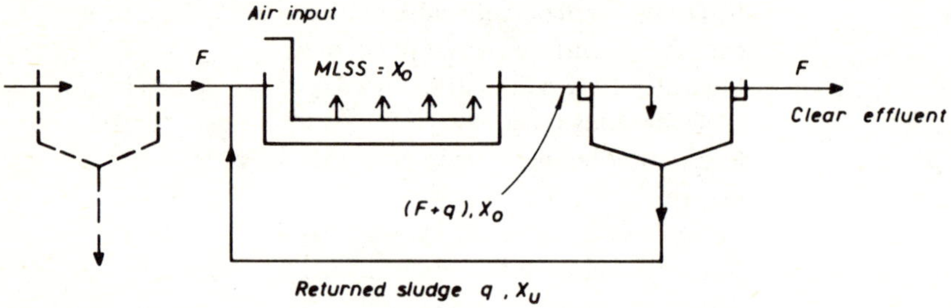

Fig. 4.1 Schematic diagram of the activated sludge process

4.1.2 Percolating filters

In this process, aerobic settled sewage is sprayed onto stones, plastic media etc. and organisms grow in a film a few millimetres thick on the media. Higher organisms such as fly larvae and microscopic worms help keep the voids clear and prevent ponding, i.e. standing water owing to void blockage. Excess film sloughs off the media continuously, but at greater rates during the spring. The film or sludge is settled out in a secondary settling tank, usually referred to as a 'humus' tank, as in Fig. 4.2. The sludge is not returned but some of the effluent may be recirculated to the 'filter' inlet. Since voids of several tens of millimetres are common, adsorption rather than filtration occurs. Single pass filtration, recirculating or alternating double filtration, high rate (plastic media) filtration are common variations with the rotating disc process representing a common small scale variation (see Chapter 8, Section 8.4.3).

4.1.3 Oxidation ponds

These are typified by rectangular ponds 1–1.5 m deep in which sludge deposits to form an anaerobically digesting sludge while above the mud aerobic organisms oxidize both the products of this digestion and the soluble and colloidal BOD. Algae photosynthetically produce oxygen to allow the oxidation to continue as

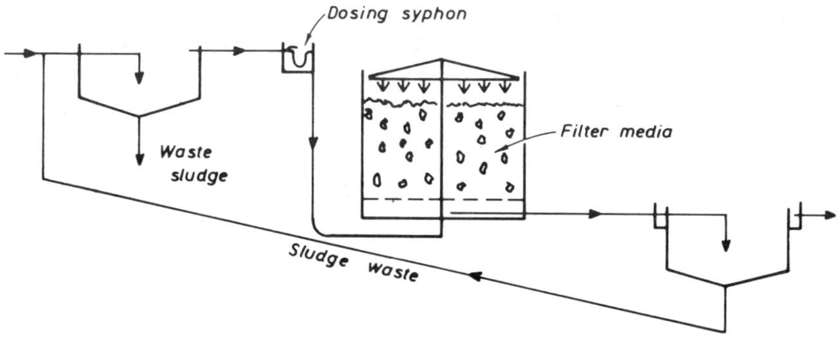

Fig. 4.2 Schematic diagram of a percolating filter system

shown schematically in Fig. 4.3. This process represents a low cost, land intensive system which has several variations including totally anaerobic ponds (for elevated temperature food processing wastes), mechanically aerated ponds and tertiary ponds which are lightly loaded in order to polish the effluent.

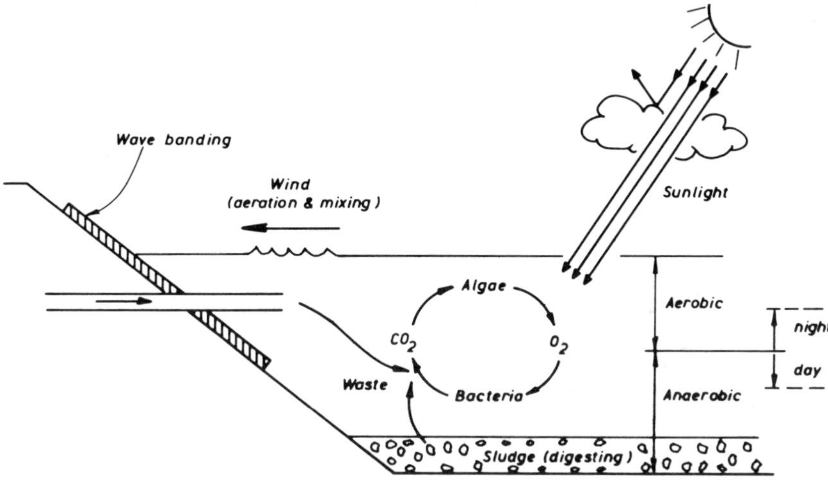

Fig. 4.3 Schematic diagram of part of an oxidation pond

4.2 Biological kinetics

The Monod equation relating the specific growth rate of bacteria to substrate (food) concentration has been shown to be applicable to sewage treatment and is given as

$$\mu = \frac{\hat{\mu} S_e}{K_s + S_e} \tag{4.1}$$

where S_e is the substrate concentration (BOD or COD in mg/l) which may be interpreted as the substrate concentration in the *outlet* stream from a mixed microbiological reactor (e.g. an extended aeration tank); K_s is a constant (mg/l) and has the value of S_e when $\mu = \hat{\mu}/2$, and $\hat{\mu}$ is the maximum value of the specific growth rate μ (time^{-1}). The specific growth rate is defined as

$$\mu = \frac{1}{X}\frac{dX}{dt} \tag{4.2}$$

where X is the biomass concentration (mg/l) and t is time.

Integration of Equation (4.2) when $\hat{\mu}$ is at its maximum value μ gives

$$\ln\frac{X}{X_0} = \hat{\mu}t \tag{4.3}$$

which can be used to give a generation time or the doubling time for a bacterial population.

Example 4.1

The value of $\hat{\mu}$ for common sewage bacteria [1] has been quoted as 0.1 h^{-1} and for the nitrifying bacteria responsible for nitrate formation as 0.08 day^{-1}. Calculate the generation time for these two types of organisms, and plot the Mönod curve for the sewage bacteria if K_s = 55 mg/l as COD. Hence find the generation time when S_e = 40 mg COD/l.

Calculation

The doubling time or generation time t_g is given by Equation (4.3) as

$$t_g = \frac{1}{\hat{\mu}}\ln\frac{2X_0}{X_0}$$

Therefore for the wastewater bacterium

$$t_g = \frac{\ln 2}{0.1} \text{ or } 6.9 \text{ h}$$

and for the nitrifying organism

$$t_g = \frac{\ln 2}{0.08} = 8.7 \text{ days}$$

If

$$\mu = \frac{\hat{\mu}S_e}{K_s + S_e}$$

then for S_e = 10 mg COD/l

$$\mu = \frac{0.1 \times 10}{55 + 10} = 0.015\ \text{h}^{-1}$$

and similarly for other values of S_e. Fig. 4.4 may be drawn from which a value of 0.042 h^{-1} for μ is found when S_e = 40 mg/l. Thus t_g at 40 mg COD/l = 16.5 h.

□ □ □

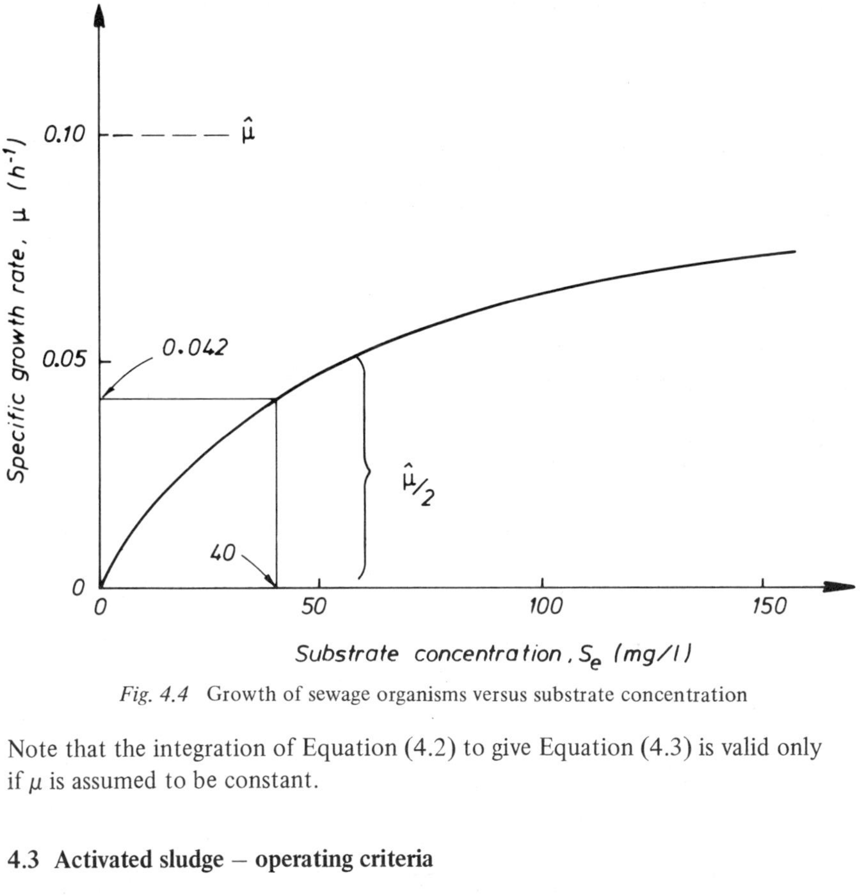

Fig. 4.4 Growth of sewage organisms versus substrate concentration

Note that the integration of Equation (4.2) to give Equation (4.3) is valid only if μ is assumed to be constant.

4.3 Activated sludge – operating criteria

4.3.1 High rate system

It can be shown from the Mönod equation that under maximum growth conditions ($\mu = \hat{\mu}$) that

$$\ln\left(1 + \frac{YL_r}{X_0}\right) = \hat{\mu}t \tag{4.4}$$

where L_r is the BOD removed (mg/l); X_0 is the mixed liquor suspended solids concentration (MLSS in mg/l) and Y is the yield of biomass dX (mg/l) for a given substrate removal dS (mg BOD/l), i.e.

$$\frac{\mathrm{d}X}{\mathrm{d}S} = -Y$$

(the negative sign implies a positive value for dX and Y when S is diminishing.) The above equations can be used to determine aeration times for high strength wastes.

Example 4.2

An activated sludge unit treats a food processing waste of 0.01 m^3/s with a BOD of 1500 mg/l for which $\hat{\mu} = 0.1\ h^{-1}$ and $Y = 0.6$. If a mixed liquor value (X_0) of 3000 mg/l is kept constant, and the returned sludge concentration is 11 000 mg/l, calculate the aeration volume required to give an effluent concentration S_e of 50 mg/l.

Calculation

In Fig. 4.1 consider the settlement tank solids balance

$$(F + q)X_0 = qX_u \tag{4.6}$$

where X_0 and X_u are the MLSS and returned sludge solids concentrations, respectively (mg/l), and F and q are the influent sewage and returned sludge flow rates, respectively (m^3/s). Thus

$$\frac{F}{q} = \frac{11\,000 - 3000}{3000} = 2.7$$

and

$$q = \frac{1}{2.7}F = \frac{0.01}{2.7} = 0.0037\ m^3/s$$

The retention time t is given by

$$t = \frac{1}{0.1}\ln\left[1 + 0.6 \times \frac{(1500 - 50)}{3000}\right]$$

$$= 2.5\ h\ \text{retention}$$

This is however a biological retention time, i.e. how long the waste must be contacted with the sludge. The gross size of the tank must accommodate a waste flow of 0.01 m^3/s plus a flow of 0.0037 m^3/s of sludge for 2.5 h. The aeration volume is thus

$$\text{tank volume} = 2.5 \times 3600 \times 0.0137 = 123 \text{ m}^3$$

□ □ □

4.3.2 Conventional system

Two equations have been presented by CIRIA[2] for cases where the BOD is in the usual range for sewage treatment, e.g. < 500 mg/l. One operates where the BOD is > 250 mg/l to give a straight line plot of BOD versus retention time (gross), e.g.

$$L_e = L_i - (250 - 9.3)\left(\frac{JtX}{2000}\right) \tag{4.7}$$

For BOD ⩽ 250 mg/l

$$L_e = 9.3 + (L_i - 9.3)\exp - \left(\frac{JtX}{2000}\right) \tag{4.8}$$

where J is a parameter defined at temperature T°C as

$$J = 0.0073\,T^2 - 0.0827\,T + 0.7162 \tag{4.9}$$

and L_e and L_i are effluent and influent BOD concentrations, respectively. This may be used to calculate the required aeration time. These equations were derived empirically to be used in a model of the activated sludge process, and as such they produce optimistically low retention times, as can be shown by reference to the later section dealing with sludge loading ratios. However, they exhibit the typical forms of empirically based equations: first a linear rate of removal of BOD with retention time when the BOD > 250 mg/l and secondly a first-order removal rate when the BOD ⩽ 250 mg/l.

Example 4.3

For an activated sludge tank treating 0.05 m^3/s of settled sewage of 275 mg/l BOD calculate the retention time and aeration tank volume required to produce an effluent of 20 mg/l when the mixed liquor is 2000 mg/l and the lowest daily temperature is 8°C.

Calculation

For the linear portion of a BOD/retention time plot, i.e. from 275 to 250 mg/l BOD

$$250 = 275 - (250 - 9.3)\left(\frac{JtX}{2000}\right)$$

$$J = 64 \times 0.0073 - 8 \times 0.0827 + 0.7162 = 0.52$$

Therefore

$$t = \frac{(250 - 275) \times 2000}{-(240.7) \times 0.52 \times 2000} = 0.20 \text{ h}$$

for 250 to 20 mg/l

$$20 = 9.3 + (250 - 9.3) \exp - \left(\frac{0.52 \times t \times 2000}{2000} \right)$$

Therefore

$$\ln (0.0445) = -0.52t$$

$$t = 6.0 \text{ h}$$

The total time is 6.0 + 0.2 = 6.2 h.

The retention time in this example is the apparent retention time, i.e. volume of the aeration tank/flow of sewage from primary settlement.

Thus the tank volume = 6.2 x 3600 x 0.05 m^3

$$= 1120 \text{ m}^3$$

□ □ □

If the activated sludge process is operated with vary long retention times in the aeration tank it is referred to as an extended aeration process. The differences between the various processes are not clearly defined and it has therefore become common, in recent years, to describe the type of activated sludge process in terms of 'sludge age' or 'sludge loading' as in Section 4.3.4.

4.3.3 Decay stage – aerobic digestion

When the food source is fully depleted the bacteria break down according to the expression

$$\frac{dX}{dt} = -K_d X \tag{4.10}$$

where K_d is an endogenous or decay rate constant.

Integration gives

$$\ln \left(\frac{X}{X_0} \right) = -K_d t \tag{4.11}$$

Because it is the organic (volatile) fraction that is decaying, it is usual to express Equation (4.11) in terms of the volatile biomasses X_v and X_{v0} and a volatile decay constant K_d'

$$\ln\left(\frac{X_v}{X_{v0}}\right) = -K_d' t \tag{4.12}$$

The process of aerobically storing sludge and allowing it to decay is called aerobic digestion and is used where anaerobic digestion is not practicable, for example, in the case of small scale units serving small populations.

Example 4.4
3 m^3 of excess activated sludge of 75% volatile matter at 8000 mg/l is aerobically digested for 10 days. The sludge is then settled to a 1.5% slurry and transported from the site; what is the transported volume if $K_d' = 0.07$ day^{-1}.

Calculation
3 m^3 at 8000 mg/l total solids ≅ 3 m^3 at 6000 mg organics/l. Therefore, after 10 days aerobic digestion, the concentration of organic material is

$$X_v = X_{v0} \exp(-K_d'\ 10) = 6000 \exp(-0.07 \times 10) = 2980 \text{ mg/l}$$

which is equivalent to a total solids concentration of 2980 + 2000 mg/l or 4980 mg/l assuming the 2000 mg/l inorganic solids remain intact.

The total mass is 4980 x 10^{-6} x 3 x 10^3 kg or 14.9 kg. At 1.5% solids (15 000 mg/l or 15 kg/m^3), the total volume transported is 14.9/15 m^3 or approximately 1 m^3.

□ □ □

Further discussion of aerobic digestion is given in Chapter 8, Section 8.3.6, since it is most commonly found as part of a small scale treatment plant.

4.3.4 Sludge age and sludge loading

The sludge age or mean cell residence time θ_c for an activated sludge unit is defined as

$$\theta_c = \frac{\text{Mass of sludge solids in the aeration tank (kg)}}{\text{Mass of sludge solids wasted daily (kg/day)}} \tag{4.13}$$

and has units of days. This may be expressed as

$$\theta_c = \frac{VX_0}{F_w X_0} \tag{4.14}$$

where V is the aeration tank volume (m^3), X_0 is the MLSS concentration and F_w the daily waste flow of excess sludge.

A common estimate of excess sludge production would be 0.6 kg dry sludge solids per kg BOD removed.

The sludge loading ratio is defined as

$$\text{SLR} = \frac{\text{Mass of BOD}_5\text{ input to aeration tank (kg/day)}}{\text{Mass of MLSS in aeration tank (kg)}} \tag{4.15}$$

and has units of day^{-1}. It may also be written

$$\text{SLR} = \frac{FL_i}{VX_0} \tag{4.16}$$

where F is the sewage inflow (m^3/day). This is also known as the food/micro-organism ratio (F/M).

Values of θ_c and SLR can readily be determined from typical plant analyses.

Example 4.5

Table 4.1 lists some operating parameters for two common types of activated sludge plant. Calculate the values of θ_c and the SLR for each case.

Table 4.1 Common activated sludge operating details*

Parameter	Conventional activated sludge (1)	Extended aeration (2)
BOD load	0.04 kg/cap . day (after settlement)	0.06 kg/cap . day
Aeration volume	50–70 l/cap . day	230 l/cap . day
Sludge wasting	0.6 kg/kg BOD removed	0.5 kg/kg BOD removed
MLSS	3000 mg/l	3000 mg/l

* see also Appendix 4.

Calculation

For conventional activated sludge, case 1.

The sludge wasted = 0.04 x 0.6 kg/cap . day

The MLSS = 50 x 3000 x 10^{-6} kg/cap. (using an aeration volume of 50 l/ 50 l/cap . day) hence

$$\theta_c = \frac{50 \times 3000 \times 10^{-6}}{0.04 \times 0.6} = 6.3 \text{ days}$$

Similarly for extended aeration, case 2.

$$\theta_c = \frac{230 \times 3000 \times 10^{-6}}{0.06 \times 0.5} = 23 \text{ days}$$

For the SLR, conventional activated sludge, case 1.

$$\text{SLR} = \frac{0.04}{0.15} = 0.27 \text{ per day}$$

and for extended aeration, case 2.

$$\text{SLR} = \frac{0.06}{0.69} = 0.09 \text{ per day}$$

□ □ □

It should be noted that in addition to the sludge actually wasted, approximately 4 g/cap . day of sludge solids may pass over the outlet weir in the effluent stream. Also the MLSS may vary from 2000 mg/l up to 6000 mg/l. Both these effects can cause considerable variations in values of θ_c.

The sludge loading ratio may be modified to become the sludge utilization rate (SUR) which is defined as

$$\text{SUR} = \frac{\text{Mass of BOD removed, kg/day}}{\text{Mass of volatile solids in reactor, kg}} \tag{4.17}$$

$$= \frac{FL_r}{VX_{v0}} \tag{4.18}$$

and has units of day^{-1}.

Often only the MLSS (X_0) is measured as opposed to the mixed liquor volatile suspended solids (X_{v0}), in which case the SUR may be expressed as $FL_r/(VX_0)$. A relationship between X_{v0} and X_0 has been given as:

$$\frac{\text{MLVSS}}{\text{MLSS}}\left(\frac{X_{v0}}{X_0}\right) = \frac{0.85}{\theta_c^{0.1}} \tag{4.19}$$

Example 4.6

Two separate activated sludge processes treat identical sewage flows of 200 l/cap . day with inlet and outlet BODs of 300 mg/l and 20 mg/l. The aeration volumes are, for plant 1; 0.070 m^3/cap . with $\theta_c = 6$ days; and for plant 2; 0.230 m^3/cap . with $\theta_c = 15$ days. If the MLSS in each case is 3000 mg/l calculate the SUR values for the two processes.

Calculation

Using Equation (4.19) the values for X_{v0} are for plant 1

$$X_{v0}(1) = 3000 \times 0.85/6^{0.1} = 2130 \text{ mg/l or } 2.13 \text{ kg/m}^3$$

and for plant 2

$$X_{v0} = 3000 \times 0.85/15^{0.1} = 1950 \text{ mg/l or } 1.95 \text{ kg/m}^3$$

The mass of BOD removed in each case is

$$\text{BOD removed} = 200\ (300 - 20) \text{ mg/cap . day} = 0.056 \text{ kg/cap . day}$$

For plant 1

$$\text{SUR (1)} = \frac{0.056}{0.07 \times 2.13} \text{ kg BOD/kg MLVSS, day}$$

$$= 0.38 \text{ day}^{-1}$$

(In common practice, settlement would reduce the 0.056 to 0.04 and SUR to 0.27 day^{-1}).

For plant 2

$$\text{SUR (2)} = \frac{0.056}{0.23 \times 1.95}$$

$$= 0.12 \text{ day}^{-1}$$

4.3.5 Sludge yield coefficient

When a mass of BOD is utilized in a set time interval ($\mathrm{d}S/\mathrm{d}t$) a mass of new cells is produced ($\mathrm{d}X_v/\mathrm{d}t$) thus

$$\frac{\mathrm{d}X_v}{\mathrm{d}t} = \frac{-Y\mathrm{d}S}{\mathrm{d}t} \tag{4.20}$$

Y is constant over a range of SUR values and is the yield coefficient (the –ve sign is included to indicate that $\mathrm{d}X_v/\mathrm{d}t$ is +ve when S is decreasing).

The net yield over a period of time is influenced by process parameters such as the respiration rate K_d and any washout and inputs.

4.3.6 Sludge growth and sludge wasting

An expression relating the sludge age θ_c in terms of the yield coefficient can be derived with the help of a diagram of an activated sludge system showing the biomass flows (Fig. 4.5). A continuity equation for the volatile component of the sludge for a completely mixed activated sludge system (CMAS) is

$$FX_{vi} + \left(\frac{\mathrm{d}X_{vo}}{\mathrm{d}t}\right)V = (F - F_w)X_{ve} + F_wX_{vw} \tag{4.21}$$

where suffixes *i*, *e* and *w* represent inlet, effluent and waste flows, respectively. $(\mathrm{d}X_{vo}/\mathrm{d}t)V$ is the cell growth in the reactor (or aeration tank).

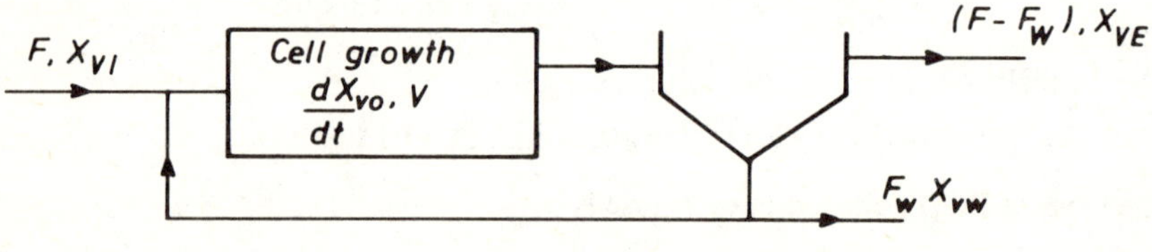

Fig. 4.5 Volatile solids in the activated sludge process

Now

$$\frac{dX_{v0}}{dt} = Y\frac{dS}{dt} - K'_d X_{v0} \tag{4.22}$$

(dS is assumed to be decreasing with time.)

Thus if X_{vi} and X_{ve} are small with respect to X_{vw} and dX_v/dt, dividing by $X_{v0}V$ gives

$$\frac{Y}{X_{v0}}\frac{dS}{dt} - K'_d = \frac{F_w X_{vw}}{VX_{v0}} \tag{4.23}$$

Now the ratio of X_{vw}/X_{v0} is the same as the ratio X_w/X_0 since the volatile fraction is similar in X_w and X_0. Thus

$$\frac{F_w X_w}{VX_0} = \frac{F_w X_{vw}}{VX_{v0}} \tag{4.24}$$

i.e.

$$\frac{1}{\theta_c} = Y \times \text{SUR} - K'_d \tag{4.25}$$

(note that dS/dt and X_v both refer to the mg of substance per unit *aeration* volume).

Values for Y and K'_d are conventionally 0.7 and 0.08 day^{-1}, respectively, thus for many plants Equation (4.25) can be rewritten as

$$\frac{W_v}{M_B} = 0.7 - 0.08\left(\frac{M_v}{M_B}\right)K \tag{4.26}$$

where W_v is the sludge to be wasted (kg/day); M_B is the total mass of BOD removed (kg/day) and M_v is the total MLVSS in the reactor or aeration tank (kg). K is a constant with value 1.0 but with dimension T^{-1} and units of day^{-1}.

Example 4.7

A wastewater treatment plant treats a per capita flow of 200 l/day from a population of 10 000. The inlet and outlet BODs are 350 mg/l and 20 mg/l, respectively. If the aeration tank has 3000 mg/l MLVSS and a capacity of 0.07 m^3/cap., calculate the volume of sludge to be wasted assuming a settled sludge concentration of 10 000 mg/l as volatile solids.

Calculation

$$M_B = 10\,000 \times 200 \times (350 - 20) \times 10^{-6} \text{ kg/day}$$
$$= 660 \text{ kg/day}$$

$$M_v = 0.07 \times 10\,000 \times 3000 \times 10^{-3} \text{ kg}$$
$$= 2100 \text{ kg}$$

Thus

$$W_v = 0.7 \times 660 - 0.08 \times 2100$$

$$= 294 \text{ kg/day of volatile solids (dry)}$$

Which, at 10 000 mg/l (1% solids) is equivalent to 29.4 m^3.

□□□

Values for Y and K_d show wide variations even in small scale experimental aeration units and they are rarely measured on large scale works. It is therefore common to find empirically based expressions for sludge wastage being used. One such expression which is commonly used in the UK [3] is

$$W_v = FL_i(0.2 + \text{SLR}^{1/2}) \tag{4.27}$$

The wide variations for sludge wastage rates often quoted in the literature are illustrated by the comparison of the value obtained by using the above equation for the conditions quoted in Example 4.7 with the value of 29.4 m^3 above.

Example 4.8

For the sludge considered in Example 4.7 calculate the volume of sludge to be wasted if the waste sludge is 75% organic using Equation (4.27).

Calculation

$$F \times L_i = 10\,000 \times 200 \times 350 \times 10^{-6} = 700 \text{ kg/day}$$

$$\text{SLR} = \frac{700}{2100/0.75} \text{ per day} = 0.25 \text{ per day}$$

Thus

$$W_v = 700(0.2 + 0.25^{1/2})$$

$$= 490 \text{ kg/day}$$

$$= 49 \text{ m}^3$$

□□□

4.4 Aeration

The input of oxygen into activated sludge plants may be carried out by mechanical aeration, compressed air diffusion, pure oxygen input or deep shaft air entrainment. The transfer of oxygen to water is, however, governed by physical laws. Therefore, much of the following is applicable not only to activated sludge units but also to stream aeration theory.

The saturation concentration of O_2 in pure water (C_s) at T°C is given (approximately) by Equation (4.28) for temperatures between 0 and 30 °C. A table of C_s values for higher temperatures and salt solutions is given in Appendix 5.

$$C_s = 14.61 - 0.3943\,T + 0.007714T^2 - 0.0000646T^3 \text{ mg/l} \qquad (4.28)$$

At 0 °C, C_s = 14.6 mg/l and at 20 °C, C_s = 9.29 mg/l, i.e. C_s decreases with increasing temperature. For sewage, C_s is approximately 85–90% of C_s for pure water.

The rate of input of air into a mass of liquid (sewage) is often more relevant than the C_s value since aeration tanks commonly operate at 10–20% saturation, i.e. 1–2 mg O_2/l, at 15 °C.

The mass of oxygen transferred per unit time $\mathrm{d}m/\mathrm{d}t$ is proportional to the difference between the actual concentration of oxygen in the liquid C and the saturation value C_s, i.e.

$$\frac{\mathrm{d}m}{\mathrm{d}t} \propto (C_s - C) \qquad (4.29)$$

Dividing by the tank volume V and using a common version of the proportionality constant gives

$$\frac{\mathrm{d}C}{\mathrm{d}t} = K_L a(C_s - C) \qquad (4.30)$$

The constant $K_L a$ has units of time^{-1} and incorporates the area of transfer, the diffusional process parameters (temperature, ionic strength etc.) and a hypothetical liquid film thickness. Equation (4.30) can be written using the dissolved oxygen deficit, D which is equal to $C_s - C$.
Since

$$D = C_s - C \qquad (4.31)$$

then

$$\mathrm{d}D = -\mathrm{d}C \qquad (4.32)$$

and

$$\frac{\mathrm{d}D}{D} = -K_L a \ \mathrm{d}t \qquad (4.33)$$

thus

$$\frac{D}{D_0} = \exp(-K_L a \times t) \qquad (4.34)$$

or

$$\ln\left(\frac{D}{D_0}\right) = -K_L at \tag{4.35}$$

if D_0 is the value of D when $t = 0$.

The above expressions are useful in river reaeration studies when $K_L a$ is replaced by K_2.

However, in the context of sewage treatment it is more useful to express the last equation as

$$\ln\left(\frac{C_s - C}{C_s - C_0}\right) = -K_L at \tag{4.36}$$

where C_0 is the value of C when $t = 0$.

Equation (4.36) can be used to describe the maximum rate at which an aerator can transfer oxygen.

Example 4.9

Two aerators are used to aerate separate, equal volumes of water which have been previously deaerated by the addition of sodium sulphite catalysed by cobalt chloride. The oxygen concentrations were read at different times using a dissolved oxygen probe. Calculate the $K_L a$ values for the two aerators and hence calculate the maximum capacity of both aerators.

Table 4.2 Results of aeration test. Water temperature is 16 °C

Time (s)	0	250	500	750	1000	1250	1500
Aerator 1 Dissolved oxygen (mg/l)	0	4.4	6.8	8.2	9.0	9.4	9.7
Aerator 2 Dissolved oxygen (mg/l)	0	2.2	3.8	5.0	6.1	7.3	7.6

Calculation

For 16 °C, $C_s = 10.0$ mg/l which is the asymptotic value in Table 4.2. For $t = 500$ s for aerator 1, the oxygen concentration is 6.8 mg/l. Thus

$$\ln\left(\frac{C_s - C}{C_s}\right) \text{ is } \ln\left(\frac{10 - 6.8}{10}\right) \text{ or } -1.14$$

Similarly the other values for $\ln[(C_s - C)/C_s]$ can be calculated and plotted against time t as in Fig. 4.6. This gives slopes of -2.3×10^{-3} s^{-1} and -9.5×10^{-4} s^{-1} for $K_L a$ for aerators 1 and 2, respectively. The maximum capacities, or the maximum rate, of input of oxygen into the solution is given by $\mathrm{d}m/V\mathrm{d}t = K_L a C_s$ (i.e. when $C = 0$). Thus capacities of 2.3×10^{-2} mg/ls and 9.5×10^{-3} mg/ls are found for aerators 1 and 2, respectively.

□□□

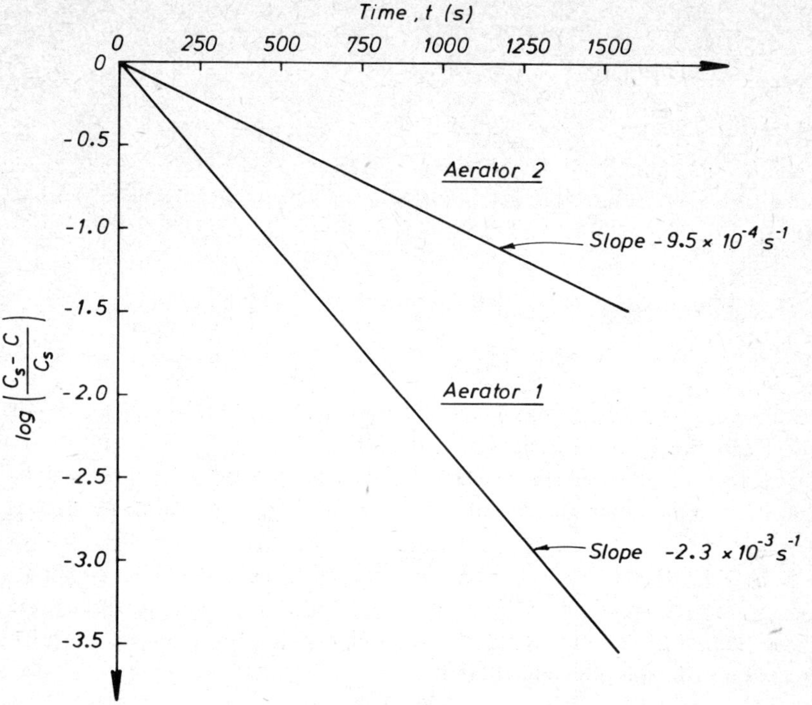

Fig. 4.6 Determination of K_La

The oxygenation capacity of a given aerator in practice is affected by many different factors, including barometric pressure changes and the presence of organics and synthetic detergents. One expression which enables an experimental value of oxygenation capacity (OC) to be related to standard conditions is

$$\mathrm{OC}_{exp} = \alpha \mathrm{OC}_{std} \left(\frac{\beta C_s - C}{C_s}\right)\left(\frac{P - P_w}{(101.3 - P_w)}\right) \tag{4.37}$$

where OC_{std} is the oxygenation capacity for pure water at standard pressure (101.3 kN/m^2 or kPa); OC_{exp} is the capacity under experimental conditions with pressure $P(kN/m^2)$; β and α are factors (usually less than unity) which allow for changes in C_s and K_La owing to the presence of organics and synthetic detergents, and P_w is the unreacting vapour pressure. Experimental results carried out in pure water can therefore be related to other fluids or suspensions such as wastewaters.

Example 4.10

An aerator delivers a maximum of 5×10^{-2} g O_2/ls in a tank containing pure water at 10 °C at a pressure of 100 kN/m^2. It is to be used to aerate the same size tank containing sewage for which $\beta = 0.9$ and $\alpha = 0.95$ at 10 °C under a pressure of 87 kN/m^2. Calculate the loss in maximum oxygenation capacity under these conditions.

Calculation

Since the maximum oxygenation is required, the expression $(\beta C_s - C)/C_s$ reduces to β since the maximum capacity occurs when $C = 0$. P_w at 10 °C is 1.2 kN/m^2 [4]. Thus the oxygenation capacity in sewage is

$$\mathrm{OC}_{sewage} = 5 \times 10^{-2} \times 0.95 \times 0.9 \left(\frac{87.0 - 1.2}{101.3 - 1.2}\right) \text{ g/ls}$$

$$= 3.7 \times 10^{-2} \text{ g/ls}$$

□□□

4.4.1 Variation of oxygen capacity with pressure

The correction of OC values for changes in pressure in the last example may be significant in mountainous countries since pressure varies with altitude Z(m) as

$$P = 101.3 \left(1 - \frac{Z\,0.13}{1000}\right) \text{ kN/m}^2 \tag{4.38}$$

The reduction from 100 to 87 kN/m^2 in the last example is therefore equivalent to a change of approximately 1000 m.

4.4.2 Variation of oxygen capacity with temperature

Increasing temperature affects the capacity

(*a*) by increasing K_La thus

$$(K_La)_T = (K_La)_{20\,°\mathrm{C}} \times 1.024^{(T-20)} \tag{4.39}$$

(*b*) by reducing C_s in accordance with Equation (4.30).

These two effects tend to cancel each other out.

Example 4.11

For the same conditions as in Example 4.10, calculate the change in oxygenation capacity for a temperature change from 10° to 25 °C.

Calculation

$$\text{Maximum OC} = \frac{\mathrm{d}m}{V\mathrm{d}t} = K_LaC_s$$

Now C_s values for 10 °C and 25 °C are 11.3 mg/l and 8.4 mg/l, respectively, or, expressed as a ratio 1.35.

Values of K_La are in the ratio $\dfrac{1.024^{(10-20)}}{1.024^{(25-20)}}$ or 0.70

Thus the ratios of the capacities at 10° and 25 °C are

$$\frac{\text{OC}\,(10°)}{\text{OC}\,(25°)} = 1.35 \times 0.70 = 0.95$$

i.e. a change of 5%. A further correction to the OC sewage value in Example 4.10 if the sewage were at 25 °C, would be to divide the value of 3.7×10^{-2} mg/ls by 0.95, i.e.

$$\text{OC sewage} = 3.9 \times 10^{-2} \text{ mg/ls.}$$

□ □ □

4.4.3 Aeration economy and efficiency

In most cases in sewage treatment the oxygen is transferred by blowing air into or entraining air in, the sewage. For both mechanical aerators and diffusers an overall efficiency of approximately 1.6 to 1.8 kg O_2/kWh (0.44 to 0.50 mg O_2/J) can be estimated. Approximate figures for oxygen requirements for various processes are given in Table 4.3.

Table 4.3 Oxygen requirements for different processes

	kg O_2/kg BOD removed
Low sludge age (partial treatment)	0.9
Conventional (20 : 30 : BOD : SS)	1.2 to 1.3
Extended aeration	2.0

Therefore, for a conventional process, an economy of approximately 1.35 kg BOD removed per kWh (0.38 mg BOD/J) is common. This includes allowance for sludge oxidation.

The efficiency of transfer of oxygen from air is usually very low implying that large volumes of air have to be entrained.

Example 4.12

Assuming an oxygen transfer efficiency of 6%, calculate the air required per capita for an activated sludge plant removing 0.04 kg BOD/cap . day, assuming 1.3 kg O_2/kg BOD removed. Also calculate the air supply per m^3 aeration capacity given a per capita aeration volume of 70 l.

Calculation

At NTP (i.e. 273 K and 101.3 kN/m^2) one gramme molecular weight (O_2 = 32 g) is contained in 22.4 l of gas. Since air contains 21% oxygen, 32 g O_2 are contained in 22.4/0.21 or 106.7 l air, or 1 g O_2 is contained in 3.33 l air. At an aeration efficiency of 6%, 3.33/0.06 or 55.6 l of air must be blown or entrained in order to transfer 1 g O_2.

Thus 0.04 kg BOD requires 1.3 x 0.04 or 0.052 kg O_2 for a conventional process, or, 0.052 kg O_2 is transferred per 52 x 55.5 l or 2.89 m^3 of air blown. This is the per capita air requirement.

The air required per m^3 of aeration capacity is

$$2.89 \times \frac{10^3}{70} m^3 \text{ or } 41.3 \text{ } m^3 \text{ air.}$$

□□□

It should be noted that for small sewage works the problems of blocking of fine diffusers leads to the use of less efficient, large orifice (> 15 mm) diameter pipes which show very low efficiencies.

4.4.4 Pure oxygen systems

The use of pure oxygen activated sludge systems is becoming increasingly popular, especially in the USA. Advantages claimed are reduced overall costs, and, more specifically, smaller aeration tanks, better sludge thickening and lower sludge production rates.

The oxygen is either produced by cryogenic air separation or pressure-swing systems. The former method involves liquifying air at −185 °C and utilizing the different distillation temperatures of the O_2 and N_2 fractions. Many different pressure-swing oxygen production techniques are possible, however, the principle involves passing high pressure air through zeolite beds which preferentially absorb nitrogen, thus producing oxygen rich gas. When the bed is saturated, the pressure is reduced and the N_2 (and some oxygen) desorbs, leaving the bed renewed. Using duplicate beds and purging the beds with air enriched with oxygen enable production rates of up to 0.8 kg O_2/kg bed material to be achieved at 90% oxygen concentration expressed as volume/volume (*v*/*v*).

Table 4.4 lists some commonly quoted operational parameters for pure oxygen systems. The aeration tanks are covered to ensure an oxygen rich atmosphere above the sewage and in a typical (Unox) system, the oxygen is fed into four completely mixed tanks operated in series. The gas feed employs a sparge tube and impeller system. Reference by the reader to a more specialist text [5] will clearly indicate some controversy regarding the more outstanding claims for pure oxygen systems and the following discussion is therefore limited to main points on which there appears to be some agreement.

(*a*) *Oxygen concentrations and oxygen transfer*

Henry's Law states that the saturation concentration of oxygen in fresh water is nearly proportional to the partial pressure of oxygen in the air above the water.

Table 4.4 Representative literature values for air and pure oxygen activated sludge plants

	Air plant	Pure oxygen plant
	Aeration stage	
MLSS (mg/l)	3300	8000
MLVSS (mg/l)	2500	6500
Aeration period (h)	4–5	1–2
Oxygen requirement (kg O_2/kg BOD removed)	1.2–1.3	1.3–1.4
DO (mg/l)	1–3	6–10
K_d (day^{-1})	0.10	0.06
Y	0.60	0.05
SLR (/day)	0.2–0.3	0.3–0.5
	Settlement tank	
Overflow rate* (m^3/m^2 day)	40–60	40–60
Mass loading (kg/m^2 day)	< 200	250
Weir loading* (m^3/m day)	< 200	150–200

The partial pressure of oxygen in the air is the product of the percentage of oxygen in the air (v/v) and the net barometric pressure $P - p_w$. This information can be used to estimate values of C_s for enriched oxygen systems and also the oxygenation capacity of diffusers using enriched oxygen.

Example 4.13
Calculate both the maximum solubility of oxygen in sewage at 10 °C and the increase in oxygenation capacity of an aerator over that for normal air, using an enriched oxygen atmosphere of 90% $O_2(v/v)$.

Calculation
From Equation (4.28), C_s at 10 °C using normal air at 101.3 kN/m^2 is 11.3 mg/l. From Henry's law

$$C_s \propto \frac{O_2\%}{100} \times (P - p_w) \tag{4.40}$$

Hence for an O_2% of 90, the value of C_s will increase to (11.3 x 0.90)/0.21 or 48.4 mg/l. Multiplying by 0.95 to correct for sewage solids gives a value for C_s of 46.0 mg/l.

The OC is equal to $K_La C_s$ from before. If K_La is assumed not to change, the improvement in OC from use of 90% O_2 enriched air is in the ratio

$$\frac{48.4}{11.3} \times OC_{(water)}$$

i.e. more than a 4 fold increase.

□□□

(*b*) *Retention times and MLSS values*
From Equation (4.25)

$$\frac{1}{\theta_c} = Y\,\mathrm{SUR} - K_d$$

or

$$\frac{1}{\theta_c} = \frac{YFL_r}{VX_{v0}} - K_d \qquad (4.41)$$

Clearly a plot of $1/\theta_c$ versus L_r/tX_{v0} will give a plot with slope Y and ordinate intercept $-K_d$. Such plots have been used to determine values of Y and K_d of 0.047 and 0.06 per day, respectively, for pure oxygen activated sludge systems compared with respective values of 0.6 and 0.10 per day for air. Table 4.4 indicates that one of the most striking differences between air and enriched oxygen units is the high MLVSS values. This gives a considerable reduction in retention time while keeping θ_c value at conventional levels.

Example 4.14
The retention time, L_r, MLSS, Y and K_d values for a conventional plant are 0.2 day, 350 mg/l, 3500 mg/l, 0.60 and 0.10 day^{-1}, respectively. The L_r, MLVSS, Y and K_d values for a pure oxygen system are 350 mg/l, 6500 mg/l, 0.46 and 0.06 day^{-1}, respectively. If the θ_c values are the same for both plants, compare the aeration tank retention times for the two processes.

Calculation
From Equation (4.25)

$$\frac{1}{\theta_c} = \frac{YL_r}{tX_{v0}} - K_d \qquad (4.42)$$

for the plant using normal air

$$\frac{1}{\theta_c} = \frac{0.6 \times 350}{0.2 \times 3500} - 0.1 = 0.2 \text{ day}^{-1}$$

and

$$\theta_c(\text{air}) = 5 \text{ days.}$$

For the pure oxygen plant

$$0.2 = \frac{0.46 \times 350}{t_{O_2}\ 6500} - 0.06$$

thus t (oxygen) = 0.095 day or 2.3 h

The aeration volume for the same flow is roughly halved in the pure oxygen process.

□ □ □

Other design criteria including those given in Table 4.4 are similar to activated sludge units using air.

4.5 Percolating filters

The simplest form of percolating filter is the single pass system where settled sewage is sprayed onto stone media after which the effluent is collected in under drains and settled to remove dead film.

A formula has been proposed [6] to relate the efficiency of BOD removal (E fraction) to operating parameters:

$$E = \frac{1}{1 + 0.44[W/(fV)]^{0.5}} \tag{4.43}$$

where W is the BOD load applied (kg/day); f is a recirculation factor, discussed in the next section, and V is the gross or overall volume of filter media (m^3). If the filter is a single pass or non-recirculating filter, $f = 1$. This can be used to calculate dimensions for a typical filter.

Example 4.15
A single pass filter is to treat 0.01 m^3/s of settled sewage of 300 mg/l BOD. Calculate the volume and area of filter media necessary to achieve 80% reduction in BOD.

Calculation

$$W = 0.01 \times 8.64 \times 10^4 \times 0.3 \text{ kg/day} = 259 \text{ kg/day}$$

$$E = 0.80$$

Therefore

$$0.80 = \frac{1}{1 + 0.44(259/V)^{0.5}}$$

or

$$V = 802 \text{ m}^3$$

An average depth of filter would be 1.8 m giving an area of 446 m^2. The load is equivalent to approximately 0.32 kg/m^3day. A check is needed to ensure that the hydraulic flow rate is within moderate range of 0.5 to 2.5 m^3/m^3day. In the above case the hydraulic loading is $(0.01 \times 8.64 \times 10^4)/802$ or 1.08 m^3/m^3day. Note that to achieve a final BOD of 45 mg/l, $V = 1610$ m^3; the BOD loading rate is 0.16 kg/m^3day.

□□□

Where the filtration scheme is designed to allow recirculation of the effluent from the secondary settlement tank, the recirculation factor f is defined by

$$f = \frac{1 + R/F}{(1 + 0.1\, R/F)^2} \tag{4.44}$$

where R and F represent the recirculated and settled sewage flows, respectively. The effect of recirculation is to reduce the volume of media required for a set efficiency.

Example 4.16

For the flow and loading stated in Example 4.15 the filter is to be designed to give a recirculated flow of 0.005 m^3/s. Calculate the reduction in volume of filter media.

Calculation

$$F = \frac{1 + 0.005/0.010}{[1 + 0.1\,(0.005/0.010)]^2} = 1.36$$

Thus

$$0.80 = \frac{1}{1 + 0.44\,[259/(V1.36)]}0.5$$

and $V = 590$ m^3 or 27% reduction in volume.

The BOD loading rate is 0.43 kg/m^3day and the hydraulic loading rate is

$$\frac{0.015 \times 8.64 \times 10^4}{590} = 2.2 \text{ m}^3/\text{m}^3\text{day}$$

□□□

Note that as the recirculation increases at lower efficiencies, the hydraulic loading rate becomes limiting since increasing this rate washes a considerable amount of solids off the filter. However, provided a secondary settlement tank is used, the majority of these solids will be retained and the BOD should show little deterioration.

An alternative expression has been proposed [7] using the hydraulic loading rate, (F/V) (m^3/m^3day); the specific surface area A_s (m^2/m^3) and the temperature T(°C), i.e.

$$L_e = L_i \exp\left[-\frac{0.037 \times 1.08^{(T-15)}}{F/V} A_s\right] \tag{4.45}$$

Example 4.17

Using the same figures as in Example 4.15 and assuming $T = 10$ °C and $A_s = 70\ \text{m}^2/\text{m}^3$, calculate the filter volume using Equation 4.45.

$$L_e = 0.2 \times 300\ \text{mg/l} = 60\ \text{mg/l}$$

Thus

$$\ln\left(\frac{60}{300}\right) = \frac{-0.037 \times 1.08^{(-5)}\ 70}{F/V}$$

and

$$F/V = 1.10,$$

hence $$V = \frac{0.010 \times 8.64 \times 10^4}{1.10} = 785\ \text{m}^3$$

Note that a rise in influent sewage temperature to 15 °C would reduce V to 537 m^3.

□□□

4.6 Large scale rotating disc unit

Rotating biological filters are examples of modifications of percolating filters which have been used for many years but have recently increased in popularity, especially for small scale use. They consist of large, 2–3 m diameter, discs of braced plastic netting, expanded polystyrene or other material rotating on a shaft at right angles to the sewage flow. They are immersed to about 40% of their diameter and alternately contact sewage then air. A microbiological film grows on the discs and oxidizes the organic matter in the sewage in a similar manner to a percolating filter. The discs are arranged into baffled groups or stages to prevent short circuiting of the flow. The disc stage is followed by a 'humus' tank in which material sloughing off the disc settles out.

The design procedures are empirical and usually are based upon a hydraulic loading rate (flow/unit disc area) for large units treating domestic wastes or on a BOD loading rate/unit disc area for small scale systems.

The procedure given below is that presented by Steels[8]. A design basis for a small disc unit is included in Chapter 8, Section 8.4.3.

Example 4.18

A disc unit is to be designed for a population of 5000 with a daily per capita water usage of 200 l. The BOD after primary settlement is 0.04 kg/cap . day and the unit is to produce an effluent of $\leqslant$ 20 mg/l BOD

Calculation

$$\text{BOD removal} = \frac{0.04 \times 10^6}{220} - 20 \text{ mg/l} = 162 \text{ mg/l}$$

$$\text{BOD percentage removal} = \frac{162}{182} = 89\% \text{ removal}$$

$$\text{BOD load} = 5000 \times 0.04 \text{ kg/day} = 200 \text{ kg BOD/day}$$

$$\text{Average flow (m}^3\text{/min)} = \frac{5 \times 10^3 \times 220 \times 10^{-3}}{24 \times 60} = 0.76 \text{ m}^3\text{/min}$$

Table 4.5 Flow correction factors for population equivalents (Courtesy I. H. Steels)

Persons equivalent	Q (m³ /min) correction factor
10 000	1
5000 to 10 000	1.1 to 1.2
1500 to 5000	1.2 to 1.3
400 to 1500	1.3 to 1.5
400	1.5

A correction factor of 1.2 is suggested (see Table 4.5), i.e.

$$\text{Average flow} = 0.76 \times 1.2 \text{ or } 0.92 \text{ m}^3\text{/min}$$

From Fig. 4.7, in order to achieve 89% removal at an inlet concentration of 182 mg BOD/l a value of A/Q of 16 800 m^2 min/m^3 is required where A is the disc area (m^2) and Q is the flow rate m^3/min (corrected). The surface area A = 16 800 x 0.92 m^2 = 15 500 m^2.

Four shafts or stages are to be used in series; this should give the required 89% BOD removal. Increasing the number of stages enables a reduction in area to be made according to Table 4.6. Thus

$$A(\text{corrected}) = 15\,500 \times 0.87 = 13\,485 \text{ m}^2$$

Assuming 3 m diameter discs are used, with a surface area of 14.1 m^2 (both sides are used) the total number of discs required is 13 485/14.1 = 956, i.e. four stages or shafts each of 239 discs.

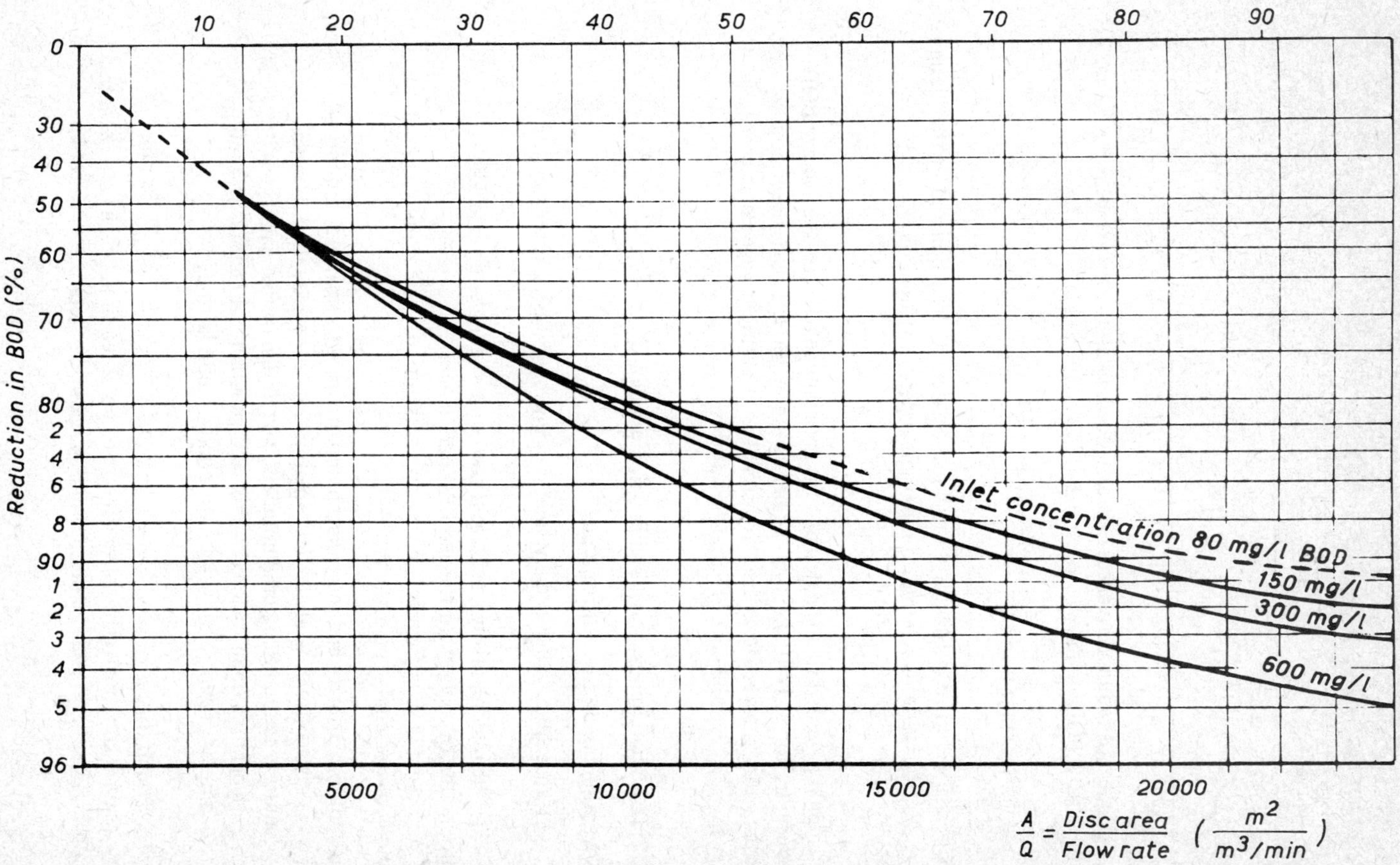

Fig. 4.7 Rotating disc efficiency curve (by the courtesy of I. H. Steels [8])

Table 4.6 Correction of A factor for stage numbers greater than two (Courtesy I. H. Steels)

Number of stages	A, correction factor	Maximum % BOD reduction
3	0.91	85
4	0.87	90
more than 4	0.85	90

A check is required to ensure that the loading on the first stage is not more than 100 g BOD/m^2 day.

$$\text{1st stage loading} = \frac{5000 \times 0.04 \times 10^3}{14.1 \times 239} = 59 \text{ g/m}^2\text{ day}$$

which is therefore satisfactory. For a more detailed discussion and further design checks, the reader is referred to the original paper by Steels [8].

□□□

It should be noted that the design basis given above relies on the average BOD concentration at DWF, i.e. 0.04 kg per capita in 200 l. This has been found to be satisfactory since rotating discs are capable of treating sewage satisfactorily while being subject to the normal variations of flow.

4.7 Oxidation ponds (waste stabilization ponds)

Oxidation ponds are of three types

(1) anaerobic, which usually treat concentrated wastes at temperatures above 15 °C

(2) facultative, involving sludge digestion in the lower layers and algal production of oxygen for the aerobic upper layers

(3) aerated ponds which are akin to a mixture of waste stabilization ponds and activated sludge plants.

In addition, primary and secondary ponds are used to treat raw and settled sewage, respectively, while tertiary (maturation) ponds are used to polish effluents by reducing bacterial numbers. The design of the main types of ponds varies depending upon socio-economic and geographical factors. The following are examples of general principles for ponds in mid latitudes.

4.7.1 Facultative ponds

A common equation relating the volume of pond required (V, m^3) to the input load (L_D, mg/day) and the mean monthly water temperature of the coldest month (T_m, °C) is given [9] as

$$V = 3.5 \times 10^{-5} \, L_D \, 1.085^{(35 - T_m)} \qquad (4.46)$$

The value of L_D is derived from the product of the contributing population, the per capita flow per day (l/cap . day) and the BOD_5 concentration (mg/l) in the case of a secondary (presettled sewage) pond. For a primary pond, the ultimate BOD is used.

Depths are often limited to 1.5 m and general dimensions of length to breadth of 1:2 to 1:3 prevail with the longest dimension aligned with the prevailing wind.

Zero outflows may occur due to losses by seepage and evaporation thus producing excessive algal crops which may produce anaerobic conditions in the top layers; this effect must be considered in the design procedure.

Example 4.19

Calculate dimensions for a primary pond to serve 1000 people discharging 200 l/cap . day and 55 g BOD_5. T_m is 12 °C and the predicted seepage, evaporation and rainfall values are 200 mm/yr, 800 mm/yr and 600 mm/yr, respectively. The BOD_5 = 70% $BOD_{ultimate}$.

Calculation

The total daily load (as BOD) is 1000 x 55 x 10^3/0.7 mg/day = 78.6 kg/day. Thus

$$V = 3.5 \times 10^{-5} \times 78.6 \times 10^6 \times 1.085^{(35 - 12)}$$

$$= 17\,960 \text{ m}^3$$

Assuming a depth of 1.5 m this is equivalent to an area of 11 970 m^2. The surface loading rate (usually expressed as BOD_5/ha . day) is

$$\frac{1000 \times 55 \times 10^{-3}}{11970 \times 10^{-4}} \text{ kg/ha day} = 46 \text{ kg BOD}_5\text{/ha . day}$$

The daily inflow is

$$Q = 200 \times 1000 \times 10^{-3} \text{ m}^3 \text{ or } 200 \text{ m}^3$$

or 17 mm/day or 508 mm/month over the total area.

Assuming constant seepage (17 mm/month), summer evaporation of twice the average rate (133 mm/month) and summer conditions of half the average rainfall rate (25 mm/month) the net outflow for the month of greatest liquid loss, would be

$$(508 + 25 - 133 - 17) \times 10^{-3} \times 11\,970 \text{ m}^3\text{/month}$$

$$= 4590 \text{ m}^3\text{/month or } 151 \text{ m}^3\text{/day}$$

It appears that zero discharge would not be a problem unless seepage were to greatly increase. In this circumstance a plastic membrane or other means of sealing would be used. The gross retention time is 17960/200 or 90 days.

The pond is small and wave band material may not be needed, the maximum water fetch (at length/breadth ratio of 1:2) being 155 m.

□□□

4.7.2 Retention time in oxidation ponds

Equation (4.46) does not, however, indicate the sensitivity of BOD removal to variations in retention time. It is possible, for a particular climate or small country, to obtain a relationship between the percent BOD_5 reduction and the retention time in the primary or secondary pond. For example, in New Zealand (mid latitude, temperate climate) a regression analysis of primary ponds[10] gave the expression relating the removal efficiency $E\%$ to the retention time t(days) as

$$E = 100 - 691\, t^{-0.9} \tag{4.47}$$

Example 4.20

A population of 10 000 discharges 200 l/cap . day domestic sewage. Calculate the area required for an oxidation pond system 1.5 m deep to achieve BOD removals of 80 and 90%.

Calculation

For 80% BOD removal, using Equation (4.47)

$$80 = 100 - 691\, t^{-0.9}$$

hence

$$\log t = \frac{-1}{0.9} \log \left(\frac{100 - 80}{691}\right)$$

and $t = 51$ days. For $E = 90\%$, $t = 111$ days. The area (A) requirement is given by

$$A = \frac{F \times t}{1.5}$$

for 80% efficiency

$$A = \frac{10\,000 \times 0.2 \times 51}{1.5} = 68\,000 \text{ m}^2$$

and at 90% efficiency

$$A = 148\,000 \text{ m}^2$$

□□□

4.7.3 Maturation ponds

These are tertiary treatment schemes used to produce a high quality effluent. Owing to the presence of algae in the effluent, the SS values may be high; coliform counts are, however, usually relatively low.

A common equation relating bacterial numbers/100 ml in the outlet stream, N_t, to influent stream numbers/100 ml, N_0, the retention time, t, days, and the breakdown coefficient of the particular organism, k, day^{-1}, is

$$\frac{N_t}{N_0} = \frac{1}{(1 + kt)} \qquad (4.48)$$

and hence for n ponds in series

$$\frac{N_t}{N_0} = \frac{1}{(1 + kt)^n} \qquad (4.49)$$

Maturation ponds are usually designed to have retention times of approximately 20 days, however, for large flow rates values of 8 to 10 days should be possible. The value of k for *E. coli* (a faecal pollution indicator organism) in the above equation has been quoted as 2 day^{-1} and for *S. typhi* as 0.8 day^{-1}.

Example 4.21

Effluent is discharged from a secondary oxidation pond at 200 m^3/day; it contains 10^6 *E. coli.*/100 ml. Calculate the maturation pond volumes required to reduce this value by 99.8% using (1) a single pond, and (2) ponds in series having retention times of 10 days and 20 days.

Calculation

For 99.8% reduction

$$\frac{N_t}{N_0} = \frac{1}{500}$$

For case 1

$$\frac{1}{500} = \frac{1}{(2t + 1)}$$

Thus t = 249 days.

For series ponds with retention time 10 days

$$\frac{1}{500} = \frac{1}{(2 \times 10 + 1)^n}$$

hence

$$n \ln 21 = \ln 500$$

and n = 2.0.

For series ponds each with retention time 20 days

$$n = 1.7$$

If two ponds are chosen for the two latter cases, the volumes required are 49 800 m^3, 4 000 m^3 and 8 000 m^3 for the single pond, two 10 day ponds and two 20 day ponds, respectively.

□□□

Questions

Question 4.1

An activated sludge unit is designed on the basis of 70 l aeration capacity/head containing 3000 mg/l MLSS of which 75% is volatile. If 0.04 kg BOD/cap . day is received by the aeration stage at 200 l/cap . day and sludge is wasted at a rate of 2.2 l/cap . day at 12 000 mg/l, calculate the sludge age, the sludge loading ratio and the sludge utilization ratio. Assume effluent SS and BOD values are 30 mg/l and 20 mg/l, respectively.

Question 4.2

The growth rate of wastewater organisms is found to vary with substrate composition as given below:

μ (h^{-1})	0.073	0.27	0.53	0.67	0.72	0.79	0.80
S_e (mg/l)	10	50	200	500	1000	10 000	100 000

Find K_s and the generation time of the bacteria (doubling time) under maximum growth conditions.

Question 4.3

An anaerobic digester stores sludge solids at 9500 mg/l as total sludge dry SS taken from a plant operating at a value of θ_c of 10 days. If K_d' for volatile solids is 0.07 day^{-1}, calculate the time for the digester contents to reduce to 6000 mg/l as dry SS (use Equation (4.19).

Question 4.4

A rotating disc plant is to be designed for a domestic population of 3000 persons each discharging 0.037 kg BOD/cap . day after primary settlement and using 250 l/cap . day. If the final effluent is to be $\leqslant$ 20 mg BOD/l calculate the number of 3 m diameter discs in each stage.

Question 4.5

A single pass percolating filter system is to be constructed to serve a population of 5000 discharging 200 l/cap . day at an average BOD of 350 mg/l. If primary settlement removes 40% BOD, using Equation (4.43) calculate the volume of filter media required to give an effluent of 35 mg/l BOD. If settled effluent is returned at a rate of 100% of the inflow to the filter calculate the new effluent BOD.

References

1 Peil, K. M. and Gandy, A. F. Jun. (1971), *Applied Microbiol.*, **21**, 253–56.
2 CIRIA (1973), Construction industry research and information report no. 46, May, UK.
3 Publication 110; *Simplex surface aeration*, Ames, Crosta, Babcock Ltd., Heywood, UK.
4 Weast, R. C. (ed.) (1974), *Handbook of Chemistry and Physics*, D-159, Chemical Rubber Co. Press, Ohio.
5 Speece, R. E. and Malina J. F. Jun. (eds.) (1973), *Application of Commercial Oxygen to Water and Wastewater Systems*, Water Resources Symposium No. 6, Centre for research in water resources, University of Texas.
6 National Research Council (1946), *Sew. Wks. J.*, **18**, 789–1028.
7 Bruce, A. M. and Merkens, J. C. (1973), *J. Inst. Wat. Poll. Centr.*, **72**, 499–527.
8 Steels, I. H. (1974), *J. Effluent and Wat. Treat.*, **14**, 431–45.
9 Gloyna, E. F. (ed.) (1971), *Waste Stabilisation Ponds*, W. H. O. monograph series No. 60, Geneva.
10 Ferrier, D. A. and Ralston, J. (1970), *N. Z. Engineering*, **25**, 263–70.

Further reading

Wastewater Treatment Plant Design (1977), Joint publication Water Pollution Control Federation (USA) and ASCE, New York, Chapters 14 and 15.
Barnes, D. and Wilson, F. (1978), *Chemistry and unit operations in sewage treatment*, Applied Science Publishers, London, Chapters 6 and 7.
Baker, J. M. and Graves, Q. B. (1968), Recent approaches to percolating filter design. *J. San. Eng. Div. Jnl. ASCE.*, **S.A.1.**, 65–84.

5
Tertiary treatment

Nomenclature

Symbol	*Definition and units*	*Dimensions*
S_I	Suspended solids in influent to process, mg/l	ML^{-3}
S_E	Suspended solids in effluent from process, mg/l	ML^{-3}
Q	Flow rate to process, usually m^3/day	L^3T^{-1}
A	Surface area of plot, clarifier bed etc., m^2	L^2
L	Tank length, m	L
OR	Overflow rate of humus tank, m^3/m^2 day	LT^{-1}
B	Tank breadth, m	L
h	Head loss through filter, m	L
l	Depth of filter, m	L
C_D	Dimensionless drag coefficient $C_D = (24/R_e) + (3/\sqrt{R_e}) + 0.34$ for spheres	–
V	Face velocity or bed loading, m/s or m^3/m^2s	LT^{-1}
ψ	Sphericity factor or shape factor being the ratio of diameter of sphere with same settling velocity (and same density) to geometric mean size of grains, dimensionless	–
g	Gravitational constant, 9.81 m/s^2	LT^{-2}
d	Diameter of bed grain, m	L
f	Porosity $= \dfrac{\text{volume of voids}}{\text{total bed volume}}$	–
R_e	Reynolds number Vd/V	–
υ	Kinematic viscosity, m^2/s	L^2T^{-1}
V_b	Face velocity (upward flow) during back washing, m^3/m^2s	LT^{-1}
V_x	Settlement velocity of bed grains, m/s	LT^{-1}
N	Number of bacteria at time t (in a given volume)	–

N_0	Initial number of bacteria (in a given volume)	–
k	Decay constant usually expressed/day (Base e)	T^{-1}
k'	Decay constant/day (Base 10)	T^{-1}
k''	Decay constant/day (Base 10)	T^{-1}
N'	Number of organisms/100 ml at time t, expressed as a statistical estimate based on a multiple analysis [5] most probable number (MPN)	L^{-3}
N'_0	Initial MPN	L^{-3}
V_L	Reactor volume (activated sludge tank volume), m^3	L^3
F	Inflow to reactor, m^3/day	L^3T^{-1}
F_W	Waste sludge flow, m^3/day. Assumed $\ll F$	L^3T^{-1}
X_N	Concentration of the critical nitrifying organisms in the reactor, mg/l	ML^{-3}
X_{NW}	Concentration of critical nitrifying organisms in the waste sludge stream	ML^{-3}
X_{NE}	Concentration of critical nitrifying organisms in the effluent stream, mg/l	ML^{-3}
X_{NO}	Influent concentration of nitrifying organisms, mg/l	ML^{-3}
X	MLSS in the reactor, mg/l	ML^{-3}
X_W	MLSS in the sludge waste stream, mg/l	ML^{-3}
X_E	MLSS in the sludge effluent stream, mg/l	ML^{-3}
$_N\mu$	Specific growth rate coefficient (nitrifiers) $_N\mu = \dfrac{1}{X_N}\dfrac{dX_N}{dt}$, per/day or per h	T^{-1}
$_NK_d$	Decay coefficient (nitrifiers), per day or per h	T^{-1}
$_N\mu_n$	Net specific growth rate coefficient for nitrifiers, per day or per h, $_N\mu_n = {_N\mu} - {_NK_d}$	T^{-1}
θ_c	Mean cell residence time or sludge age $\theta_c = \dfrac{\Sigma \text{ MLSS in reactor, kg}}{\text{Cell wastage, kg/day}}$	T
FMR	Food to micro-organism ratio; $\text{FMR} = \dfrac{\text{BOD input, kg/day}}{\Sigma \text{ MLSS in reactor, kg}}$ i.e. SLR (sludge loading ratio)	T^{-1}

5.1 Introduction

Tertiary treatment is an ill defined group of processes which normally occur at the end of a wastewater treatment scheme. It includes processes which seek to reduce the concentrations of nutrients, suspended solids and pathogenic organisms in the effluent stream. Several of these processes are based on empirical flow or loading rates (e.g. grass plots loaded at 0.5 m^3/m^2 day) while others are related to theoretical kinetic descriptions (e.g. assumed first-order kinetics of bacterial decay.)

5.2 Solids removal

Several tertiary treatment processes are designed to improve the effluent from a secondary treatment process by removing suspended material and with it some of the BOD. Examples of such tertiary systems are microstrainers, upward flow clarifiers, grass plots and sand filters. In most cases the design parameters are chosen empirically with little theoretical guidance being given to those responsible for installation. The following discussion is intended to give a brief description of the processes, some common design loadings and relevant comments.

5.2.1 Microstrainers

These consist of finely woven stainless steel mesh with up to 200 apertures/mm. The mesh is usually held in a circular cage form with the effluent being filtered under a head of 150 mm. Washing of the drum is continuously carried out by overhead jets of effluent.

Common operating parameters are given in Table 5.1.

Table 5.1 Microstrainer operational data

		35 μm mesh	23 μm mesh
Loading rate (m^3/m^2 day)		550	350
Head loss (mm)		150	150
Removal %SS		50–65	60–80
%BOD		25–40	35–70
Peripheral speed (m/s)		0.5	0.5
Approximate effluent standard achieved	BOD	15	10
	SS	15	10

(a) Performance

An approximate relationship between influent SS, S_I(mg/l) and effluent SS S_E(mg/l) has been proposed[1], i.e.

$$S_E = 0.11 \times S_I + 7.0 \text{ mg/l} \qquad (5.1)$$

for a 35 μm mesh strainer (125 apertures/mm^2) within the range of influent concentrations from 20 to 120 mg/l SS

5.2.2 Grass plots

These consist of plots of grassland[2] of 1 in 60 to 1 in 120 slope with the effluent being spread at a rate of approximately 0.4–0.7 m^3/m^2 day. Reductions of 50% in both SS and BOD appear to be common with reductions of 60% ammoniacal nitrogen and 30% nitrate nitrogen also being found.

Example 5.1

A flow of 200 m^3/day treated on grass plots could use three plots, two of which operate at any time, leaving one section free for cutting every 2 months, and regrading and scraping, to remove humus solids, every 5–10 years. At 0.6 m^3/m^2 day the total area is 1.5 x (200/0.6) = 500 m^2 in total or 170 m^2 per plot.

□ □ □

5.2.3 Pebble bed clarifiers

Clarifiers are usually tanks in which the effluent flows upwards through a 150 mm deep bed of pea gravel (5–10 mm diameter) at rates of up to 20 m^3/m^2 day. The removal mechanism is not by straining of solids but involves coagulation, flocculation and other mechanisms. Increasing flow rate, reduction in bed depth and increasing size of gravel all reduce the efficiency. The gravel bed requires cleaning at weekly intervals, usually by drawing down the clarifier TWL (top water level) below the bed and raking and spraying the gravel with clarified effluent. This limits the size of media and the thickness of the bed. Other materials such as wedge wire screens are also used and the tank itself may be the actual humus tank rather than a separate installation.

Percentage removals for 5–10 mm diameter gravel 150 mm deep have been shown to be approximately linear[1] below 57% removal, i.e.

$$\left(\frac{S_I - S_E}{S_I}\right) 100 = -0.96Q/A + 63 \tag{5.2}$$

where Q/A is the loading rate, i.e. inflow/surface area (m^3/m^2 day).

A typical small scale design system for a combined humus tank and clarifier is given below.

Example 5.2

Calculate the approximate dimensions for a combined humus tank and clarifier for a population of 50 persons to achieve a 15 mg/l SS effluent standard assuming conventional humus tank effluent to be 30 mg/l SS and the per capita water usage to be 180 l/day.

Calculation

The design procedure adopted is similar to that of CP 302[3], although other standards exist.

CP 302 states the capacity of a humus tank should be

$$C = (30\,P + 1500)\,\mathrm{l} \tag{5.3}$$

where P is the population and C is the tank volume (l), i.e.

$$C = 3000\,\mathrm{l} \text{ or } 3\ \mathrm{m}^3$$

Suggested tank dimensions are 1.2 m deep[3] and length L of three times the breadth B. Therefore

$$3B^2 = \frac{3.0}{1.2}$$

$$B = 0.91\ \mathrm{m};\ L = 2.74\ \mathrm{m}$$

The average surface overflow rate (OR) at 180 l/cap . day will be

$$\mathrm{OR} = \frac{50 \times 180 \times 10^{-3}}{2.49}$$

i.e. 3.6 $\mathrm{m^3/m^2}$ day, which is extremely conservative.

Using Equation (5.2) the required removal is [(30 – 15)/30] 100% or 50%, i.e.

$$Q/A = -\ \frac{50 - 63}{0.96} = 13.5\ \mathrm{m^3/m^2}\ \mathrm{day}$$

which gives a clarifier bed area of $(50 \times 180 \times 10^{-3})/13.5$ or 0.67 m^2.

For a tank breadth of 0.91 m this gives a clarifier of 0.73 m x 0.91 m x 150 m deep using 5–10 mm diameter gravel.

Alternatively, since small works suffer large fluctuations in flow rates, CP 302 suggests a maximum flow rate through the bed of 1.0 $\mathrm{m^3/m^2h}$ at 3 DWF. In this case this would imply a bed area of $(3 \times 180 \times 10^{-3} \times 50)/(24 \times 1.0)$ or 1.13 m^2 or a bed of 1.24 x 0.91 x 0.150 mm deep. In view of the small population and the consequent large flow variations, this latter size of bed is more satisfactory.

5.2.4 Sand filters

Two types of sand filters have traditionally been used both for water and waste wastewater treatment.

(a) Slow sand filters

These consist of underdrains covered by 250 mm of coarse gravel or clinker graded from 50 mm diameter at the bottom to 20 mm diameter at the top with a further 450 mm deep layer of fine (0.5–0.8 mm diameter) sand. The bed is loaded at rates up to 3.5 m^3/m^2day with effluent from secondary settlement processes. The flow is diverted to a second filter when the head loss reaches 0.6 m. After draining, the top 50 mm of sand and biological sludge are removed. The bed is remade when the sand depth reduces to 200 mm.

The design criteria in this case involve assessing the maximum flow (e.g. 3 DWF) and using a maximum flow rate of 0.1–0.15 m^3/m^2h to determine the bed area. The bed is duplicated to allow for cleaning periods. Removal efficiencies better than 60% SS and 40% BOD are usual. It is stressed that the slow sand filter involves a complex series of actions including precipitation of hydroxides as a surface film, algal production of oxygen and bacterial breakdown (oxidation) of organics.

(*b*) *Rapid gravity filters*

These consist of deeper layers (up to 1.2 m deep) of coarse (0.6–1.5 mm diameter) sand overlaying gravel. They are loaded at rates up to 230 m^3/m^2day and remove better than 70% SS and 50% BOD.

The filtration is continued until a head loss of 2–2.5 m is reached when the bed is backwashed and air scoured. The actual head loss in a rapid gravity filter is given as

$$h = 1.07\, l \frac{C_D V^2}{\psi g d f^4} \tag{5.4}$$

where h is the head loss (m) per unit filter depth l (m); C_D is the drag coefficient where $C_D = (24/R_e) + (3/\sqrt{R_e}) + 0.34$; R_e being the dimensionless Reynolds number Vd/υ; d is the particle diameter (m) and ψ is the dimensionless shape or sphericity factor (< 1); V is the approach velocity (m^3/m^2s); g is the gravitational constant (m/s^2) and f is the porosity, i.e. the ratio of (volume of voids)/(total bed volume).

The above expression can be used to determine the head loss in a simple case as follows.

Example.5.3

A rapid gravity filter consists of a 0.85 m deep bed of spherical ($\psi = 1.0$), monosized sand grains of 0.6 mm diameter. It is loaded at 1.1×10^{-3} m^3/m^2s with a suspension for which the kinematic viscosity $\upsilon = 1.2 \times 10^{-6}$ m^2/s. Calculate the head loss when the porosity is 0.35.

Calculation

$$R_e = \frac{Vd}{\upsilon} = \frac{1.1 \times 10^{-3}}{1.2 \times 10^{-6}} \times 0.6 \times 10^{-3} = 0.55$$

and

$$C_D = \frac{24}{0.55} + \frac{3}{\sqrt{R_e}} + 0.34 = 48.0$$

thus

$$h = \frac{0.85 \times 1.07 \times 48.0 \times (1.1 \times 10^{-3})^2}{1.0 \times 9.81 \times 0.6 \times 10^{-3} \times (0.35)^4} = 0.60 \text{ m}$$

□□□

When the filter is backwashed, the bed expands according to the relationship:

$$\frac{l_{expanded}}{l_{settled}} = \frac{1 - f}{1 - (V_b/V_x)0.22} \tag{5.5}$$

where $l_{expanded}$ is the expanded bed depth (m); V_b is the face velocity during backwashing ($m^3/m^2 s$) and V_x is the settlement velocity of the bed particle.

Example 5.4
A bed of settled depth 0.8 m and porosity 0.44 is backwashed at 1.5×10^{-2} $m^3/m^2 s$ ($\upsilon = 1.2 \times 10^{-6}$ m^2/s). If the sand grains are of 0.8 mm diameter and of density 2500 kg/m^3, calculate the height of the expanded bed.

Calculation
An iterative calculation of the type detailed in Chapter 2 section 2.5.1 gives a settlement velocity of 0.13 m/s for the bed particle. The bed height is therefore

$$l_{expanded} = 0.8 \frac{(1 - 0.44)}{1 - \left(\frac{0.015}{0.13}\right)^{0.22}} = 1.18 \text{ m}$$

□□□

(*c*) *Multi media filters*
In practice the bed of a rapid gravity filter is composed of different sized particles which settle after backwashing to give a graded bed with the finer particles on top. This is an inefficient system and has led to the use of multi media filters which consist of layers of particles of different densities with the largest and least dense particles on top and the smallest and densest particles on the bottom.

They are operated in a downward flow system and backwashed in an upward flow regime. The criterion for choice among the different filter media is that no particle from a lower bed shall have a settling velocity less than any particle in a higher bed. A design example is given below for the case of a three media filter using sand, anthracite and plastic (e.g. polypropylene) grains.

Example 5.5

Typical settlement curves of sand of density 2600 kg/m^3; anthracite of density 1400 kg/m^3 and plastic of density 1050 kg/m^3 are as given in Fig. 5.1.

Available anthracite has a grain size between 1.4 mm and 1.7 mm[4], i.e. a range of settling velocities (assuming spherical particles) of 0.11 to 0.09 m/s when $\upsilon = 1.2 \times 10^{-6}$ m^2/s. Plastic particles less than 6.5 mm diameter would have a settlement velocity less than the smallest anthracite particles. In practice, plastic media of 2.0 to 3.0 mm diameter having settlement velocities of 0.033 to 0.040 m/s would be suitable. Sand of greater than 0.65 mm diameter would have velocities greater than the largest anthracite grains, so sand of perhaps 0.7 to 0.9 mm diameter, with settlement velocities of 0.12 to 0.15 m/s would be satisfactory to form a lower bed. The situation would then be as in Table 5.2.

In practice the anthracite would have a shape factor of perhaps 0.75 which would have a significant effect on the fall velocity[4].

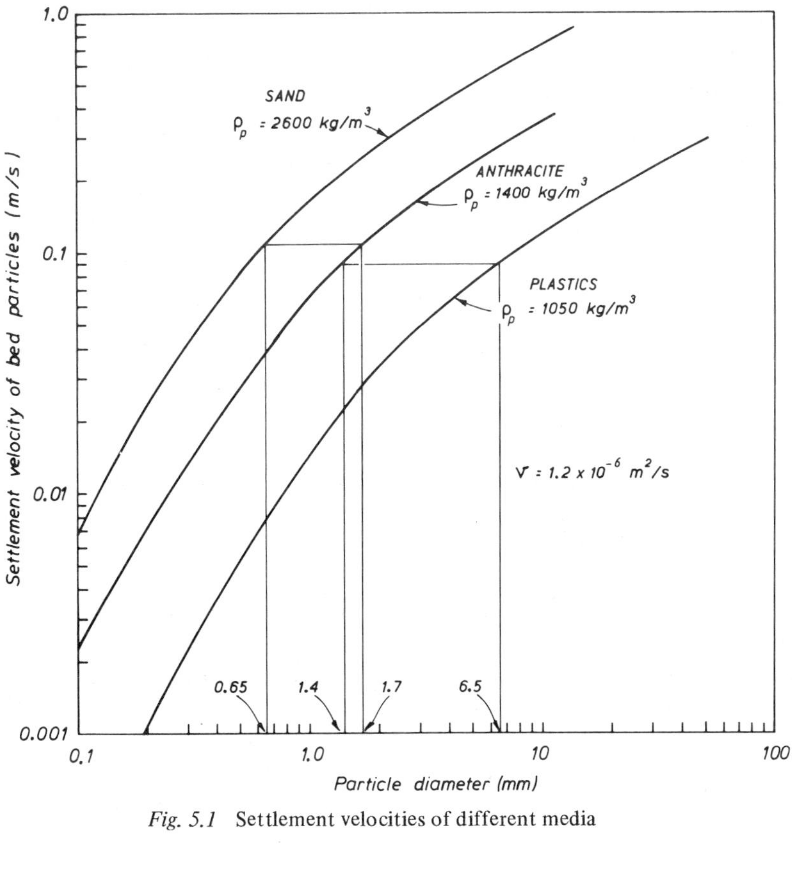

Fig. 5.1 Settlement velocities of different media

Table 5.2 Multi media filter

	Material	Density (kg/m³)	Size range (mm)	Settlement velocity* (m/s)
Upper medium	plastic	1050	2.0–3.0	0.033–0.05
Middle medium	anthracite	1400	1.4–1.7	0.11–0.09
Lower medium	sand	2600	0.7–0.9	0.12–0.15

* Assuming spherical particles and $\upsilon = 1.2 \times 10^{-6}\ m^2/s$

□□□

5.3 Disinfection

Many countries have water quality legislation which specifies limits on bacterial species such as *Erischeria coliform* and *faecal coliform*, (commonly referred to as *E. coli* and *f. coli*) in effluents discharged to receiving waters. Since domestic wastewaters commonly have 10^7 total coliforms/100 ml, which is reduced by 99% in conventional sewage treatment processes, the final effluent may still contain 10^5 total coliforms/100 ml. Typical legislation for bathing waters would require (after dilution) less than 200 *f. coli*/100 ml in the median of multiple samples, this value being perhaps 1/3 to 1/6 of the total coliform count.

Many of the common tertiary treatment processes such as slow sand filters, rapid gravity filters, grass plots and clarifiers can remove large numbers of bacteria, removals of 40–70%; 32%, 87–97% and 38%, respectively, being quoted[1].

When the physical methods are insufficient it is common to use chlorine, or, less usually, ozone, as disinfecting agents. Chlorine reacts in water to produce hypochlorous acid which dissociates to give hypochlorite ions and protons, i.e.

$$HOCl \rightleftharpoons H^+ + OCl^- \qquad (5.6)$$

At 10 °C the dissociation constant is approximately 2×10^{-8}, i.e.

$$2 \times 10^{-8} = \frac{[H^+]\,[OCl^-]}{[HOCl]} \qquad (5.7)$$

where the brackets indicate concentrations in gramme ions or g molecules/l. Since $pH = -\log_{10}[H^+]$, the above relationship can be used to predict the ratio of hypochlorite ion (OCl^-), to hypochlorous acid (HOCl) at any pH.

Example 5.6

Calculate the ratio of OCl^-/HOCl in pure water at a temperature of 10 °C and pHs of 6, 8 and 10.

Calculation
A pH of 6 means that $[H^+] = 10^{-6}$ g ions/1. Therefore

$$\frac{[OCl^-]}{[HOCl]} = \frac{2.0 \times 10^{-8}}{10^{-6}} = 0.02$$

i.e. 0.02/1.02 x 100% or 2% of the hypochlorous acid has dissociated.

At pH 8

$$\frac{[OCl^-]}{[HOCl]} = 2$$

i.e. 67% is as OCl^-, and at pH 10, 99.5% exists as OCl^-.

□ □ □

Several methods have been used to determine the rate of kill of bacteria by chlorine. The simplest is a first-order reaction, i.e.

$$\frac{dN}{dt} = -kN \qquad (5.8)$$

where N is the number of organisms at time t and k is a constant. Hence

$$\frac{N}{N_0} = e^{-kt} \qquad (5.9)$$

or

$$\frac{N}{N_0} = 10^{-k't} \qquad (5.10)$$

where N_0 is the number when $t = 0$, and k' is the constant k taken to base 10.

Chlorine reacts with organic compounds and with ammonia to produce a wide range of chlorinated organics together with mono-, di- and trichloramines (NH_2Cl, $NHCl_2$ and NCl_3, respectively). These are designated 'combined' chlorine compounds to distinguish them from the 'freely available' chlorine compounds (HOCl and OCl^-). The term residual, e.g. 'residual free chlorine', refers to that remaining after initial oxidation reactions have taken place. Compounds have widely differing reaction rates and often show deviation from Equation (5.10). A better fit to the experimental data is given by

$$\log_{10}\left(\frac{N}{N_0}\right) = -k''t^2 \qquad (5.11)$$

where k'' is a constant which varies with concentration of the chlorine compound.

Example 5.7
In clear water at 0.1 mg/l of free residual chlorine the value of k'' for *E. coli.* is

0.025 when t is in minutes. Calculate the contact time required to reduce the coliform concentration to 0.1%.

Calculation
From Equation (5.11)

$$t = \sqrt{\left(\frac{-1}{k''} \log_{10} \frac{N}{N_0}\right)}$$

$$\frac{N}{N_0} = 0.001$$

Therefore

$$t = \sqrt{\left(\frac{-1}{0.025} \times (-3.00)\right)}$$

thus $t = 11$ min.

□□□

An empirical equation has been derived for *E. coli.* (the most commonly used bacterial indicator species) which relates chlorine residual concentration C(mg/l) to contact time for *wastewaters*

$$\frac{N'}{N'_0} = (1 + 0.23\, Ct)^{-3} \tag{5.12}$$

The values of N' and N'_0 used are most probable numbers (MPNs) and represent a statistical value for the concentration of *E. coli.* determined by an MPN test [5]. The chlorine concentrations required are much higher than for clean water.

Example 5.8
Determine the likely kill of *E. coli.* in a sewage effluent ($N'_0 = 10^5$/100 ml) for a residual chlorine concentration of 2 mg/l and a contact time of 10 min.

Calculation

$$N' = N'_0\,(1 + 0.23 \times 2.0 \times 10)^{-3}$$

$$N' = 5.7 \times 10^2/100 \text{ ml}$$

or 99.4% kill.

For units treating less than 100 m^3/h, common dosages and retention times are 10 mg/l of sodium hypochlorite solution (as Cl_2) and 30 min, respectively.

□□□

The reduction of bacterial numbers in oxidation ponds is discussed in Chapter 4, Section 4.7.3.

5.4 Nutrient removal

The amount of nitrogen and phosphorus which is discharged in sewage can have a considerable effect on the growth of algae in lakes and receiving waters.

One equation for algal growth is

$$106CO_2 + 16NO_3^- + HPO_4^{2-} + 122H_2O + 18H^+ \xrightarrow{\text{(light and trace elements)}}$$

$$C_{106}H_{263}O_{110}N_{16}P_1 \text{ (algal synthesis)} + 138O_2 \qquad (5.13)$$

Example 5.9

A population of 10 000 discharges 200 l/cap . day of effluent containing 15 mg/l $N - NO_3$ and 5 mg/l phosphorus. Calculate the total algal dry weight which could be produced per day, assuming 100% efficiency of nutrient utilization.

Calculation

From the formula $C_{106}H_{263}O_{110}N_{16}P_1$ the molecular weight is

$$(12 \times 106 + 1 \times 263 + 16 \times 110 + 14 \times 16 + 31) = 3550$$

i.e. (16 x 14) kg N or 224 kg N produces 3550 kg algae and (1 x 31) kg P or 31 kg P produces 3550 kg algae.

The population produces $10^4 \times 200 \times 15 \times 10^{-6}$ kg N/day or 30 kg N/day and $10^4 \times 200 \times 5 \times 10^{-6}$ kg P/day or 10 kg P/day. The N could produce (3550/224) x 30 kg algae/day, i.e. 475 kg/day, whereas the P could produce (3550/31) x 10 kg algae/day, i.e. 1145 kg/day.

Assuming these are the sole nutrient sources there are sufficient nutrients to produce 475 kg algae/day.

□ □ □

The growth and subsequent removal of algae from a receiving water pond is one method of removal of nutrients, but note that in the above case the stoichiometry is unbalanced, i.e. P is present in excess of its stoichiometric requirement and cannot be fully removed unless some other source of nitrogen exists.

Phosphorus is more usually removed by precipitation with lime as the variable composition calcium hydroxyapetite. The method is complex and requires experimental work to determine optimum dosages, etc. [6].

Nitrogen may be removed in several ways; one of the commonest being bacterial nitrification followed by bacterial denitrification. This effectively converts organic N eventually to N_2 gas:

$$NH_4^+ + \tfrac{3}{2}O_2 \longrightarrow NO_2^- + 2H^+ + H_2O \qquad (5.14)$$

$$NO_2^- + \tfrac{1}{2}O_2 \longrightarrow NO_3^- \qquad (5.15)$$

$$NO_3^- + \tfrac{1}{3}CH_3OH \longrightarrow NO_2^- + \tfrac{1}{3}CO_2 + \tfrac{2}{3}H_2O \qquad (5.16)$$

$$NO_2^- + \tfrac{1}{2}CH_3OH \longrightarrow \tfrac{1}{2}N_2(gas) + \tfrac{1}{2}CO_2 + \tfrac{1}{2}H_2O + OH^- \qquad (5.17)$$

written using methanol as the carbon source.

A system therefore might incorporate an aerobic nitrifying stage followed by an anoxic denitrifying stage and since an anoxic effluent discharge may be unacceptable, a final aerobic phase may be required.

5.4.1 Nitrification

This step can be written from Equations (5.14) and (5.15).

$$2NH_4^+ + 3O_2 \longrightarrow 2NO_2^- + 2H_2O + 4H^+$$

$$28 \text{ g N} + 96 \text{ g } O_2$$

$$2NO_2^- + O_2 \longrightarrow 2NO_3^-$$

$$28 \text{ g N} + 32 \text{ g } O_2$$

Therefore 1 g of ammonia nitrogen requires (96 + 32)/(2 x 14) or 4.6 g O_2.

In Chapter 4 it was shown that metabolization of 1 kg BOD_5/day required an input of approximately 1.2 kg O_2/day in the activated sludge plant operating under conventional conditions. For an extended aeration plant operating under a low FMR (BOD input, kg/day/MLSS in reactor, kg) loading (typically <0.1/day) there will be a considerable additional oxygen requirement for the nitrification step.

Example 5.10

Analysis of raw sewage showed 40 mg/l organic and ammonia nitrogen associated with 200 mg/l BOD_5. Assuming 100% of the nitrogen is oxidized to NO_3^- in an extended aeration plant, calculate the O_2 requirement expressed per kg BOD_5/day, given that 1 kg BOD_5 requires 1.2 kg oxygen.

Calculation

200 mg BOD_5 from above requires $200 \times 10^{-6} \times 1.2$ or 2.4×10^{-4} kg O_2. Conversion of 40 mg NH_3–N to NO_3 requires $4.6 \times 40 \times 10^{-6}$ kg O_2 or 1.84×10^{-4} kg O_2 (from Equations (5.14) and (5.15)).

Thus, in total, 200 mg BOD_5 plus the associated organic nitrogen requires 4.2×10^{-4} mg/O_2. Therefore 1 kg BOD_5 plus the associated organic nitrogen together would require $5 \times 10^3 \times 4.2 \times 10^{-4}$ or 2.1 kg O_2/kg BOD_5.

In practice a value of 2.0–2.5 kg O_2/kg BOD_5 is used for extended aeration plants.

□ □ □

5.4.2 Nitrification kinetics

Considering the activated sludge system in Fig. 5.2 for a reactor volume V_L(m^3), inflow and waste sludge flow F and F_W, respectively (m^3/day), the rate of

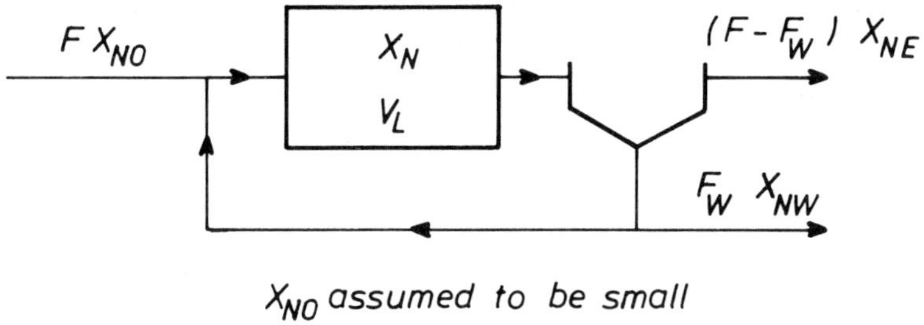

Fig. 5.2 Nitrifying organisms balance (activated sludge system)

increase in nitrifying organism concentrations X_N (mg/l) is given in terms of the specific growth rate ${}_N\mu$ and the decay rate constant ${}_NK_d$ (both referring to nitrifying organisms) as

$$\frac{dX_N}{dt}V_L = ({}_N\mu - {}_NK_d)V_LX_N - F_WX_{NW} - (F - F_W)X_{NE} \tag{5.18}$$

where X_{NE} and X_{NW} are the effluent and sludge waste stream concentrations of nitrifying organisms. For the organisms to increase

$${}_N\mu - {}_NK_d \geqslant \frac{F_WX_{NW} + (F - F_W)X_{NE}}{V_LX_N}$$

i.e.

$$\frac{1}{\text{mean cell residence time}} \tag{5.19}$$

or

$${}_N\mu - {}_NK_d \geqslant \frac{1}{\theta_c} \tag{5.20}$$

since

$$\frac{F_WX_{NW}}{V_LX_N}$$

can be assumed to be the same as

$$\frac{F_W X_W + (F - F_W)X_E}{V_L X}$$

where X_W, X_E and X are the waste sludge, effluent and reactor concentration of sludge solids. The expression $V_L X/[F_W X_W + (F - F_W)X_E]$ is defined as $1/\theta_c$ where θ_c is the sludge retention time or sludge age as in Chapter 4.

${}_N\mu - {}_N K_d$ may be written as the net specific growth rate for nitrifying organisms ${}_N\mu_n$. Since ${}_N\mu$ has a maximum value for the $NO_2^- \longrightarrow NO_3^-$ reaction carried out by nitrobacter organisms of < 0.14/day it is likely that a realistic value for ${}_N\mu_n$ of less than 0.1/day is achieved in practice. Thus the sludge age would have to be $\geqslant 10$ days to allow nitrifying organisms to increase.

Conditions which are conducive to high nitrification in the activated sludge process include:

(i) dissolved oxygen concentration > 2 mg/l
(ii) sludge age $\theta \geqslant 20$ days
(iii) FMR < 0.1
(iv) MLSS high (> 3000 mg/l)

Air supply rates, aeration volume and sludge wastage rates can be calculated as follows:

Example 5.11
A small unit treating the waste from 250 persons producing 200 l/cap . day and 0.5 kg BOD cap . day is 3 m deep. Calculate the air required per kg BOD, the aeration volume and the daily sludge wastage rate. The MLSS is to be 4000 mg/l, the settled sludge return is at 10 000 mg/l and $\theta = 20$ days.

Calculation
Since nitrification is intended, a minimum of 2 kg O_2/kg BOD is required. A 3 m deep tank would probably give an oxygen transfer efficiency of approximately 4%. Therefore

$$\frac{2000}{32} \times \frac{22.4}{0.04} \times \frac{10^{-3}}{0.21}\ \text{m}^3 \text{ air/kg BOD}$$

or 167 m^3 air/kg BOD is required, assuming 22.4 l air at NTP contains 0.21 x 32 g O_2, see Chapter 4, Section 4.4.

The total air requirement is therefore

$$250 \times 0.05 \times 167\ \text{m}^3/\text{day or } 2090\ \text{m}^3/\text{day}$$

Assume the FMR is to be 0.05/day, i.e. MLSS = (0.05 x 250)/0.05 = 250 kg MLSS.

At an average MLSS concentration of 4000 mg/l in the aeration tank this gives a volume requirement of $(250/0.004) \times 10^{-3}$ m^3 or 62.5 m^3 aeration capacity.

Since θ is to be 20 days, the sludge wasted/day should be 1/20 the total sludge in the reactor. However, an inadvertent waste of MLSS occurs in the effluent. If this is estimated as 20 mg/l dry SS the mass of SS lost per day is $20 \times 10^{-6} \times 250 \times 200$ kg/day or 1 kg MLSS/day if the value of F is assumed to be very large compared with F_W. The sludge wastage rate in the waste stream should therefore be:

$$\frac{250}{20} - 1 \text{ kg/day}$$

i.e. 11.5 kg MLSS/day.

If this is removed from the secondary settlement tank, in order to keep the volume to a minimum, then the volume to be wasted, assuming a settled sludge concentration of 10 000 mg/l, is

$$\frac{11.5}{10\,000 \times 10^{-6}} \text{ l/day}$$

or 1.15 m^3/day.

□ □ □

5.4.3 Alkalinity changes in nitrification and denitrification

From the equations for nitrification and denitrification it can be seen that the nitrification stage produces acidity as H^+ ions, and the denitrification stage produces alkalinity as OH^- ions. The alkalinity of waters is usually referred to in terms of mg/l of $CaCO_3$ equivalent. Thus for the nitrification step:

$$N \longrightarrow NO_3^- + 2H^+$$

$$14\text{ g} \qquad\qquad 2\text{gH}^+ \tag{5.21}$$

Now 2 g H^+ is equivalent to 100 g of $CaCO_3$ since we can write a reaction such as:

$$2HCl + CaCO_3 \longrightarrow CaCl_2 + CO_2 + H_2O \tag{5.22}$$

$$2\text{ g H}^+ + 100\text{ g CaCO}_3$$

Thus 1 g N which is oxidized reduces the alkalinity by 100/14 or 7.1 g $CaCO_3$.

For the denitrification step:

$$NO_3^- \longrightarrow N + OH^- \tag{5.23}$$

and since 17 g $OH^- \equiv 1$ g $H^+ \equiv 50$ g $CaCO_3$, then denitrification of 1 g of N as NO_3^- produces 3.6 g of alkalinity expressed as $CaCO_3$. Since waters in many parts of the world have low alkalinity, there is a danger that the alkalinity may be removed by nitrification and the pH will drop below the point where bacteria cease to be operative.

Example 5.12

A wastewater plant is operated in a nitrifying phase in order to convert 40 mg/l of oxidizable N to NO_3^-. If the water contains only 200 mg/l of alkalinity (as $CaCO_3$) calculate the amount of lime which must be added per litre if a minimum alkalinity of 100 mg/l (as $CaCO_3$) is to remain at all times.

Calculation

1 mg/l oxidizable N uses up 7.1 mg/l $CaCO_3$. Hence 40 mg/l would require 284 mg/l $CaCO_3$. Allowing for the safety factor of 100 mg/l $CaCO_3$ residual, a concentration of 184 mg/l $CaCO_3$ alkalinity is needed.

$$100 \text{ mg/l } CaCO_3 = 76 \text{ mg/l } Ca(OH)_2$$

(slaked lime)

i.e. 184 x (76/100) or 140 mg of slaked lime would be required per litre of wastewater.

□□□

5.4.4 Single vessel nitrification – denitrification

If the nitrification and denitrification reactions are carried out in the same tank, the net change in alkalinity is halved, e.g. 40 mg/l oxidizable N now requires only 142 mg/l $CaCO_3$ alkalinity, and the water may be able to cope with this demand. Such units are common in Australia and operate on a cycle of

(1) aeration causing carbonaceous oxidation and nitrification
(2) settlement and anoxic denitrification
(3) decantation of the top liquor as effluent
(4) cycle restarts as in (1).

Because small schemes using this intermittent system avoid separate storage by allowing the sewage to continually enter the vessel at any point in the cycle they are referred to as 'continuous intermittent' plants! Fig. 5.3 gives typical values for the cycle when nitrification and subsequent denitrification is used.

(a) Denitrification

The denitrification rate is usually expressed as the mg NO_3^- removed per g MLSS hour, and can be used to determine the time required for denitrification in large scale works. The calculation assumes that after nitrification, in a single vessel nitrification–denitrification system, the sludge settles and becomes anoxic, thus allowing denitrification to proceed.

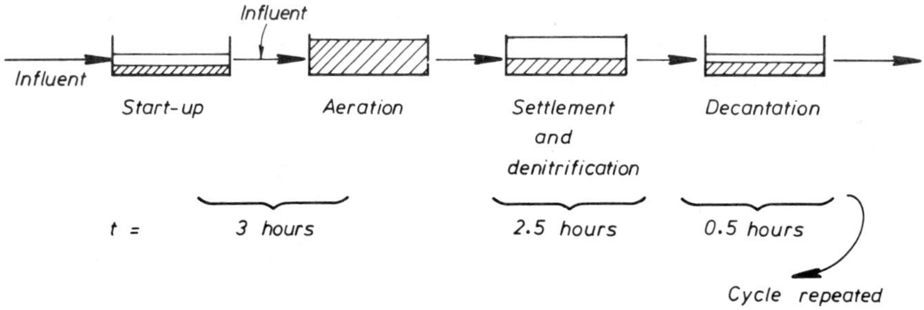

Fig. 5.3 Single vessel nitrification–denitrification cycle

Example 5.13
A single vessel activated sludge unit is run on a denitrifying cycle. The sludge has an oxygen uptake rate of 10 mg O_2/g MLSS h and at the end of the aerobic phase the dissolved oxygen concentration is 4 mg O_2/l. The MLSS concentration is 6000 mg/l and the denitrification rate is 0.5 mg NO_3^-–N/g MLSS h in anoxic conditions. Calculate a time interval for the denitrification phase if the NO_3^-–N concentration is 15 mg/l at the end of the aerobic phase.

Calculation
The deoxygenation rate is

$$10 \times 6000 \times 10^{-3} \text{ mg } O_2/\text{l h}$$

$$= 60 \text{ mg } O_2/\text{l h}$$

After aeration ceases, the time to remove 4 mg O_2/l would be

$$\frac{4}{60} \text{ h or 4 min}$$

The rate of denitrification for an MLSS of 6000 mg/l is

$$0.5 \times 6000 \times 10^{-3} \text{ mg } NO_3^-\text{–N/l h}$$

$$= 3 \text{ mg } NO_3^-\text{–N/l h}$$

The time required to remove 15 mg N is therefore 15/3 h or 5 h plus the 4 min required for the onset of anoxic conditions.

□ □ □

It is possible to achieve greater rates of denitrification than that quoted in the last example by using methanol as the organic substrate. In this case, a MLSS of 3000 mg/l is usually used in the reactor and a separate clarifier stage is necessary.

Example 5.14

A waste contains 16 mg/l NO_3–N and is to be denitrified in an anoxic reactor containing 3000 mg/l MLSS. The denitrification rate using methanol as the organic source is 7 mg N/g MLSS h. Calculate the theoretical retention time in the reactor and the methanol requirement.

Calculation

The rate of denitrification is 7 x 3.0 mg NO_3–N/h, i.e. the theoretical retention time is 16/21 h or 0.8 h. (in practice 2 h would be used).

From Equations (5.16) and (5.17) 14 g NO_3–N requires 5/6 the molecular weight of methanol for denitrification, i.e.

$$\tfrac{5}{6}(12 + 3 + 16 + 1) \text{ or } 27 \text{ g } CH_3OH$$

Therefore the methanol requirement is (16 x 27)/14 mg methanol per litre of waste.

Questions

Question 5.1

A rapid gravity filter 1 m deep is composed of monosized spherical sand grains 1 mm diameter (*d*) with a clean bed porosity of 0.4. The filter is loaded at 5×10^{-3} m^3/m^2s with an aqueous suspension of 40 mg/l of suspended solids of density 1500 kg/m^3. Assuming 100% filtration efficiency and uniform filtration throughout the bed calculate the length of filter run (in days) if the final head loss is 3 m, given

$$\mu = 1.31 \times 10^{-3} \text{ N s/m}^2$$

$$C_D = \frac{24}{R_e} + \frac{3}{R_e^{1/2}} + 0.34$$

$$R_e = \frac{\rho V d}{\mu}$$

$$h_f = 1.067 \frac{C_D V^2 l}{\psi g d f^4}$$

Question 5.2

A seaside town discharges 0.2 m^3/s of sewage effluent with an MPN of 1.5×10^6 *E. coliforms*/100 ml. Statutory bathing limits specify an average MPN of 200 *E. coliforms*/100 ml. Assuming a dilution of 100:1 of sea water (MPN = 50/100 ml) to sewage effluent, calculate the daily mass of chlorine required if the contact time is to be 10 min.

Question 5.3

A pebble bed clarifier operates on a present peak flow of 0.05 l/s containing 40 mg/l SS. The present effluent concentration averages 20 mg/l SS. The flow rate is to be increased to 0.07 l/s and the inlet SS is expected also to increase to 50 mg/l SS. Calculate the expected new outlet concentration.

References

1 Truesdale, G. A. (1967), Tertiary treatment for sewage works effluents. *J. Wat. Poll. Control,* **66**, 371.

2 Daviss, M. R. V. (1957), Treatment of humus tank effluent on grass plots. *Surveyor Munic. City Engr.,* 8 June, 613–14.

3 BSCP 302 (1972), *Small Sewage Treatment Works*, BSI London.

4 Ives, K. J. (1977), Deep bed filters, in *Solid/Liquid Separation Equipment Scale-up* (ed. D. B. Purchas), Uplands Press.

5 *Standard Methods for the Examination of Water and Wastewater* (1976), 14th Edn, Joint Publication: American PHA; American WWA; American Water Pollution Control Federation, Washington DC, p. 923.

6 Culp, R. L. and Culp, G. L. (1971), *Advanced Wastewater Treatment*, Van Nostrand-Rheinhold, New York.

Further reading

Barnes, D. and Wilson, F. (1978), Tertiary Treatment, in *Chemistry and Unit Operations in Sewage Treatment*, Applied Science Publishers, London.

6

Sludge treatment and disposal

Nomenclature

Symbol	*Definition and units*	*Dimensions*
MC	Moisture content. Mass of liquid in sludge/mass of sludge, usually %	–
SC	Solids content. Mass of solids (dry mass) in sludge/mass of sludge, % or fraction as stated	–
V	Volume of sludge or filtrate, m^3	L^3
M_s	Mass of dry solids in sludge, kg	M
M_w	Mass of liquid in sludge, kg	M
VM	Volatile matter (= organic content) %	–
TS	Total solids, kg	M
X	Volume of digester gas used in boiler, m^3	L^3
r	Specific resistance to dewaterability, m/kg	LM^{-1}
C	Solids concentration in sludge, kg/m^3	ML^{-3}
θ	Filtration time, s	T
μ	Absolute viscosity of filtrate, Ns/m^2	$ML^{-1}T^{-1}$
P	Pressure (difference) across filter, N/m^2	$ML^{-1}T^{-2}$
A	Filter area, m^2	L^2
R_m	Resistance of filter paper, cloth or septum/m	L^{-1}
b	Slope of θ/V versus V plot (s/m^6)	TL^{-6}
r_0	Specific resistance constant, units indeterminate	–
S	Coefficient of compressibility, dimensionless	–
C_l	Mass of solids/unit volume of liquid in the sludge, kg/m^3	ML^{-3}
Y	Yield of vacuum filter, kg/m^2s	$ML^{-2}T^{-1}$
A_f	Area of cake formation as fraction of total area	–
θ_R	Time of 1 revolution, s	T

f_c	Cake correction factor, ratio $\dfrac{\text{mass of liquid/unit mass of sludge}}{\text{mass of filtrate/unit mass sludge dewatered}}$	–
SC (Sl)	Solids content of wet sludge (fraction)	–
SC (C)	Solids content of sludge as it emerges from the filter bath (fraction)	–
T	Time of pressure filtration, h	T
k	Constant of proportionality, empirical $kg^2/m^2 s$ for T in s	$M^2 L^{-2} T^{-1}$
SC (F)	Final cake solids content at end of filter pressing, fraction	–
H	Heat energy required to raise water to incinerator exit temperature, J	ML^2T^{-2}
T_I	Inlet temperature of sludge, °C	°C
T_E	Exit temperature of exhaust gases, °C	°C
K	Calorific value or energy/volume of combustible material, J/kg	L^2T^{-2}
CM	Combustible material (almost equivalent to organic material)	–
Z	Fraction of calorific value taken up by incinerator heat loss	–

6.1 Sludge volume and moisture content

Sludges produced in wastewater treatment plants may be detritus sludges from grit removal processes, primary settlement sludges, secondary settlement sludges or occasionally tertiary treatment sludges. The obnoxious, putrescent nature of many sludges, coupled with their low dewaterability, makes disposal difficult. The magnitude of the problem is indicated by the simple calculation that, if a city of 1 million people produces waste sludge at 0.09 kg/cap . day at 97% moisture content, the city is faced with the daily disposal of 3000 m^3 of sludge.

Sludges produced in wastewater treatment typically contain very small amounts of solids distributed throughout a large volume of liquid. The moisture content (MC) is expressed as

$$\text{MC} = \left(\frac{\text{mass of liquid in the sludge}}{\text{mass of sludge}}\right) \times 100\% \tag{6.1}$$

and the solids content SC = 100 – MC (6.2)

At high MC values, the volume V_2 produced by dewatering a volume of sludge V_1, from MC_1 to MC_2 is given by the approximation

$$V_2 = V_1 \frac{(100 - MC_1)}{(100 - MC_2)} \text{ or } V_1 \frac{SC_1}{SC_2} \tag{6.3}$$

In practice most interest is shown in volume changes caused by dewatering, since small changes in MC produce large volume changes.

Example 6.1

A sludge of 99.5% MC containing 5 kg dry SS is dewatered progressively to 99, 97, 90 and 70% MC. Calculate the change in volume at each MC using Equation (6.3). Assuming that dry sludge solids and water have densities of 1400 and 1000 kg/m^3, respectively, estimate the error produced in the calculation of sludge volume at 70% MC by using the approximate equation.

Calculation

At 99.5 % MC 5 kg dry SS are associated with 1000 kg sludge. The volume is, (assuming a simple additive process)

$$V = \frac{5}{1400} + \frac{(1000 - 5)}{1000} = 0.999 \text{ m}^3\text{; or 1 m}^3 \text{ sludge}$$

Equation (6.3) gives the following volumes

$$\text{at MC 99\% : } V_2 = 1 \times \frac{(100 - 99.5)}{(100 - 99)} = 0.500 \text{ m}^3$$

at MC 97% : 0.167 m^3

MC 90% : 0.050 m^3

MC 70% : 0.0167 m^3

Using a density approach, at 99% MC 5 kg dry SS is associated with 500 kg sludge or 495 kg water. Thus the volume is

$$V_{99\%} = \frac{5}{1400} + \frac{495}{1000} = 0.499 \text{ m}^3$$

The volumes are similarly 97% MC; 0.165 m^3; 90% MC; 0.0486 m^3; 70% MC, 0.0152 m^3. Assuming the density calculations to be correct, the sludge volume at 70% MC using Equation (6.3) is approximately 10% greater than the correct volume.

□ □ □

6.1.1 Liquor production from treatment processes

A simple way of calculating the liquor produced from particular treatment processes is sometimes useful and can be estimated as follows. Since the solids content

$$SC(\%) = \left(\frac{M_s}{M_s + M_w}\right) \times 100\% \tag{6.4}$$

where M_s and M_w are the mass of solids and water, respectively, in the sludge, then:

$$(100 - SC)M_s = SC\, M_w \tag{6.5}$$

or, considering 1 kg of solids

$$\frac{100}{SC} - 1 = M_w \tag{6.6}$$

For most calculations M_w(kg) x 10^{-3} gives the associated liquor volume in m^3. The liquor released in a process which dewaters sludge from SC(1), %, to SC(2), %, is therefore

$$\text{liquor volume} \cong \frac{1}{10}\left(\frac{1}{SC(1)} - \frac{1}{SC(2)}\right) M_s\ (m^3) \tag{6.7}$$

Example 6.2
A sludge which is initially at 6% SC is dewatered in a filter press to 40% SC. Calculate the liquor produced in pressing 100 kg dry sludge solids.

Calculation
At 6% MC, 1 kg solids is associated with

$$\frac{100}{6} - 1 \text{ or } 15.7 \text{ kg } H_2O$$

at 40% SC, 1 kg solids is associated with

$$\frac{100}{40} - 1 = 1.5 \text{ kg } H_2O$$

Therefore, pressing 100 kg sludge would produce

$$100 \times (15.7 - 1.5) \text{ kg } H_2O$$

i.e. approximately 1.42 m^3 liquor.

□ □ □

6.2 Anaerobic digestion

The characteristics of wastewater sludges, and more particularly primary settlement sludges, which most severely limit the methods that may be used for

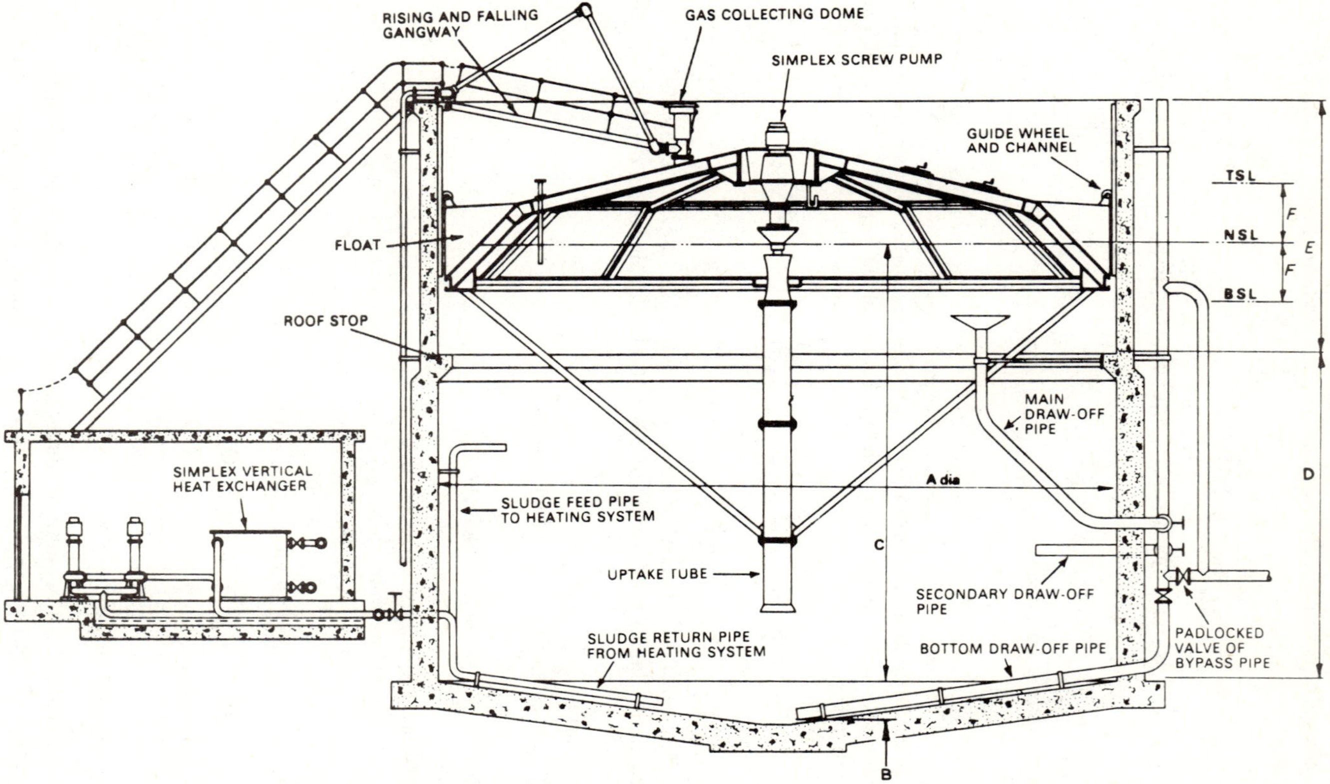

Fig. 6.1 Primary anaerobic digester with floating roof (by courtesy of Ames, Crosta, Babcock Ltd[1])

their disposal, are their smell and unaesthetic appearance. Aerobic oxidation is not usually used for other than secondary activated sludges, however, if other sludges are kept anaerobic, they will break down to produce a tarry smelling, blackish, relatively innoccuous material.

The anaerobic digestion of sludges may be carried out at cryophilic (10–15 °C), mesophilic (30–34 °C) or thermophilic (50–55 °C) temperatures. The mesophilic process is often used to produce a less obnoxious sludge and a combustible gas mixture (66% CH_4; 34% CO_2). The process is often carried out in a heated, partially mixed anaerobic primary digester of the type shown in Fig. 6.1 with a secondary dewatering tank being used prior to drying beds or final sludge disposal. Typical primary digester dimensions[1] are given in Table 6.1. Theoretical analysis of the system is difficult since in practice most digesters are only partially mixed reactors, however, general working parameters which have been found useful in practice are given in Table 6.2.

Table 6.1 Typical dimensions for Ames, Crosta, Babcock floating roof plant [courtesy of Ames, Crosta, Babcock[1]

Nominal capacity (m^3)	*A (m)	B (m)	C (m)	D (m)	E (m)	F† (m)	Size of screw pump (m)
710	10.7	0.69	7.9	5.6	5.2	1.07	0.30
990	12.2	0.76	8.5	6.0	5.2	1.22	0.46
1420	13.7	0.91	9.5	6.9	5.5	1.22	0.46
1840	15.2	0.99	10.0	7.2	6.1	1.37	0.61
2410	16.8	1.07	10.8	8.1	6.7	1.37	0.61
3540	18.3	1.22	13.3	10.5	6.7	1.52	0.61
4250	19.8	1.30	13.6	10.4	6.7	1.83	0.61

* Letters refer to Fig. 6.1

† 2 x F is the total roof movement providing a variable capacity equivalent to 7 days sludge production

Table 6.2 Conventional operating values for anaerobic digestion parameters

Primary digester volatile solids loading	1–2 kg VM/m^3 day
Solids loading	1.5–2.1 kg SS/m^3 day
Input sludge solids	2–6% SC
Total solids destruction	30–35% of TS input
Input sludge VM	70–80% VM as total solids
Volatile matter (VM) destruction	40–46% of VM input
Gas production	0.9–1.2 m^3/kg VM destroyed
Sludge gas calorific value	22 400 kJ/m^3

6.2.1 General relationships

The data given in Table 6.2 may be used to assess retention times, per capita digestion capacity, per capita gas production etc.

Example 6.3

Assuming sludge production is 0.090 kg dry SS/cap . day at 96% MC, calculate the normal retention time in a mesophilic digester using solid and volatile solids loadings from Table 6.2.

Calculation

Using a VM loading of 1.5 kg VM/m^3day at 75% VM this is equivalent to

$$\frac{0.090 \times 0.75}{1.5} \text{ m}^3/\text{cap.} = 0.045 \text{ m}^3/\text{cap.}$$

The volume of sludge produced per capita day at an average input SC of 4% (40 kg/m^3) is

$$\frac{0.090}{40} \text{ or } 2.25 \times 10^{-3} \text{ m}^3$$

The nominal retention time is therefore

$$\frac{0.045}{2.25 \times 10^{-3}} = 20 \text{ days}$$

Using a dry SS loading of 1.8 kg/m^3day gives a per capita volume of 0.09/1.8 m^3 or 0.05 m^3 and a retention time of $0.05/2.25 \times 10^{-3}$ days or 22 days.

□□□

Gas production can similarly be estimated.

Example 6.4

Determine the gas production per head of population and determine the maximum energy available given a sludge gas calorific value of 22 400 kJ/m^3.

Calculation

Assume 0.085 kg dry SS/cap . day at 75% VM with 43% VM reduction. The organics destroyed total

$$0.085 \times 0.75 \times 0.43 = 0.027 \text{ kg/cap . day}$$

At 1.1 m^3gas/kg VM destroyed, the gas volume produced is 0.030 m^3gas/cap . day, i.e.

$$22\,400 \times 0.030 \text{ kJ/cap . day or } 672 \text{ kJ/cap . day}$$

□□□

6.2.2 Heat budget calculations

If, for any reason, the temperature of the digester drops, the methane production tends to diminish. A drop in temperature usually leads to lower daily methane production and a lower $CH_4 : CO_2$ ratio in the gas. Ideally a digester should produce an excess of gas energy over that necessary to keep the digester temperature constant. Operational variables have a considerable influence on whether or not this is possible. Heat budget calculations are therefore useful both in determining those operating variables which exert most influence on digester operation and in allowing calculations to be made of the most efficient way of using the gas produced.

There are two major heat requirements in anerobic digestion which must be satisfied if the digester is to remain at the operational temperature. They are the heat

(a) to raise the daily sludge input volume to the operating temperature
(b) to keep the whole digester volume at the operating temperature.

This latter heat requirement can be calculated. However, it is more common to estimate it in terms of 'equivalent drop in temperature of digester contents' over a day. This is usually estimated from experience[1] and is often more accurate than lengthy calculations of heat transmittance through walls, roofs, etc. There is a gain in heat from the gas.

Example 6.5

Calculate the heat budget for a mesophilic digester system, serving 100 000 population, producing 0.08 kg dry SS/cap . day at 97% MC with a mean yearly temperature of 12 °C. The retention time is 25 days at 32 °C and the digester heat loss equivalent is 0.5 °C/day.

Calculation

The daily sludge mass produced = 100 000 x 0.080 or 8000 kg dry SS/day. Since 97% MC is 3% SC or approximately 30 kg dry SS/m^3 sludge, the daily volume produced is 8000/30 or 267 m^3.

The digester (liquid) volume is 267 x 25 = 6680 m^3 (probably divided equally between two digester tanks).

□ □ □

(*a*) *Heat losses*

The specific heat, or specific heat capacity of sludge at 97% is almost equal to 4.18×10^3 kJ/m^3 °C. Therefore the heat loss to the surroundings at an equivalent loss of 0.5 °C/day is

$$4.18 \times 10^3 \times 0.5 \times 6680 = 14.0 \times 10^6 \text{ kJ/day}$$

The heat required to raise a daily input of 267 m^3 of sludge through (32–12) °C is

$$4.18 \times 10^3 \times (32 - 12) \times 267 = 22.3 \times 10^6 \text{ kJ/day}$$

The total heat required per day is 36.3×10^6 kJ.

(*b*) *Gas production*
Assuming 30% total solids digestion producing 1.1 m^3 gas/kg solids destroyed, the gas production is

$$8000 \times 0.3 \times 1.1 \text{ m}^3 = 2640 \text{ m}^3 \text{ gas/day}$$

This has a theoretical heat energy of 2640 x 22 400 kJ/day or 59.1×10^6 kJ/day. There is apparently a considerable excess of heat, i.e. 22.9×10^6 kJ/day.

(*c*) *Gas utilization*
The gas is usually split between a boiler/heat exchanger system and a dual fuel engine. The former usually gives around 60–80% efficiency as heat for the digester while the latter can give 20–25% as exhaust jacket heat (to heat the digester) and 30% as usable work (driving pumps or an alternator set). The most efficient use of the gas is one where the sum of the heats produced just equals the digester requirements.

Example 6.6
Given the efficiencies quoted above and the gas production from Example 6.5 calculate the most economical gas distribution between the boiler and the dual fuel engine.

Calculation
For the 2640 m^3 of gas produced in Example 6.5 let X m^3 go to the boiler at 70% efficiency and $(2640 - X)$ m^3 to the dual fuel engine at 25% (thermal) efficiency, then

$$0.7 \times X \times 22\,400 \text{ kJ} + 0.25 \times (2640 - X) \times 22\,400 \text{ kJ}$$

should sum to give 36.3×10^6 kJ.

Thus $X = 2135$ m^3/day, i.e. 2135 m^3 gas/day is used by the boiler and 505 m^3/day is used by the dual fuel engine. The 505 m^3 of gas to the dual fuel engine therefore gives (at 30% efficiency) 505 x 0.3 x 22 400 kJ/day or 0.94×10^3 kWh/day.

□ □ □

In practice, since methane does not ignite by compression, about 7–10% of the gas energy to the engine is added as diesel fuel.

6.2.3 Effect of variables on the heat budget

One of the most readily variable parameters is the input sludge solids content. Because this has a profound influence on the amount of liquid to be sustained at the operating temperature it changes the heat budget considerably.

Example 6.7

Assuming the values of the digestion parameters given in Example 6.5 remain the same, with the sole exception of the input sludge concentration, calculate the effect on the heat budget of input sludge consolidation to 5% SC.

Calculation

The daily sludge input volume changes to 8000/50 m^3/day or 160 m^3/day. The heat requirements then become

(*a*) input sludge heat requirement $= 14.0 \times 10^6 \times \frac{160}{267}$ kJ/day

or 8.4×10^6 kJ/day

and

(*b*) the digester heat loss to the surroundings

$= 22.3 \times 10^6 \times \frac{160}{267}$ kJ/day

or 13.4×10^6 kJ/day

and the total heat requirement drops to 21.8×10^6 kJ/day, i.e. a reduction of 40%.

□ □ □

Changes in retention time have two effects; firstly reducing the heat loss to the surroundings, and secondly, reducing the total gas production.

Example 6.8

Assuming the conditions given originally in Example 6.5, consider the effects of reducing the retention time to 20 days (with a corresponding lowering of solids digestion to 28% as total solids).

Calculation

The input sludge requirement remains at 22.3×10^6 kJ/day; the ambient loss reduces to $14.0 \times 10^6 \times (20/25)$ kJ/day or 11.2×10^6 kJ/day, and the gas evolution becomes

$$2640 \times \frac{28}{30} = 2464 \text{ m}^3$$

which is equivalent to 55.2 x 10^6 kJ/day. Overall there is a change of [(55.2 – 59.1) + (14.0 – 11.1)] x 10^6 kJ/day, i.e. 1.0 x 10^6 kJ/day loss of available heat energy.

□□□

Present day practice is to insulate the digesters to cut down the ambient heat loss by up to 50%.

6.3 Specific resistance

Wastewater sludges are often difficult to dewater and even spadeable sludges may still be at 75–80% MC. Because the volume is critically dependent on the MC it is necessary to have some quantitative assessment of the ease with which a sludge can be dewatered. The most commonly accepted and theoretically useful concept is that of specific resistance [2, 3]. The specific resistance to dewatering r (m/kg) is related to volume of filtrate discharged V (m^3) in time θ (s) as

$$\frac{\theta}{V} = \frac{\mu r C}{2PA^2} V + \frac{\mu R_m}{AP} \tag{6.8}$$

where μ is the absolute viscosity (Ns/m^2); P is the pressure difference over the filter (N/m^2 or Pa); A is the filtration area (m^2); R_m is the resistance of the filter media (paper, cloth etc., /m) and C is the sludge concentration (kg/m^3).

The determination of r may be carried out in an apparatus similar to that in Fig. 6.2. Plotting values of θ/V versus V gives straight lines whose slope b is related to r as

$$r = \frac{b2PA^2}{\mu C} \tag{6.9}$$

Example 6.9

The following set of results in Table 6.3, were obtained from a specific resistance test carried out in the laboratory. Determine the specific resistance of the sludge.

Table 6.3 Results of a specific resistance test

Time	145	310	570	1440	2360	3360	4880	6170
Filtrate (m^3 x 10^{-6})	5	10	15	25	33	40	48	54

Sludge MC = 96%

Manometer pressure reading = 675 mm Hg

Buchner funnel diameter = 50 mm

Filtrate temperature = 20 °C

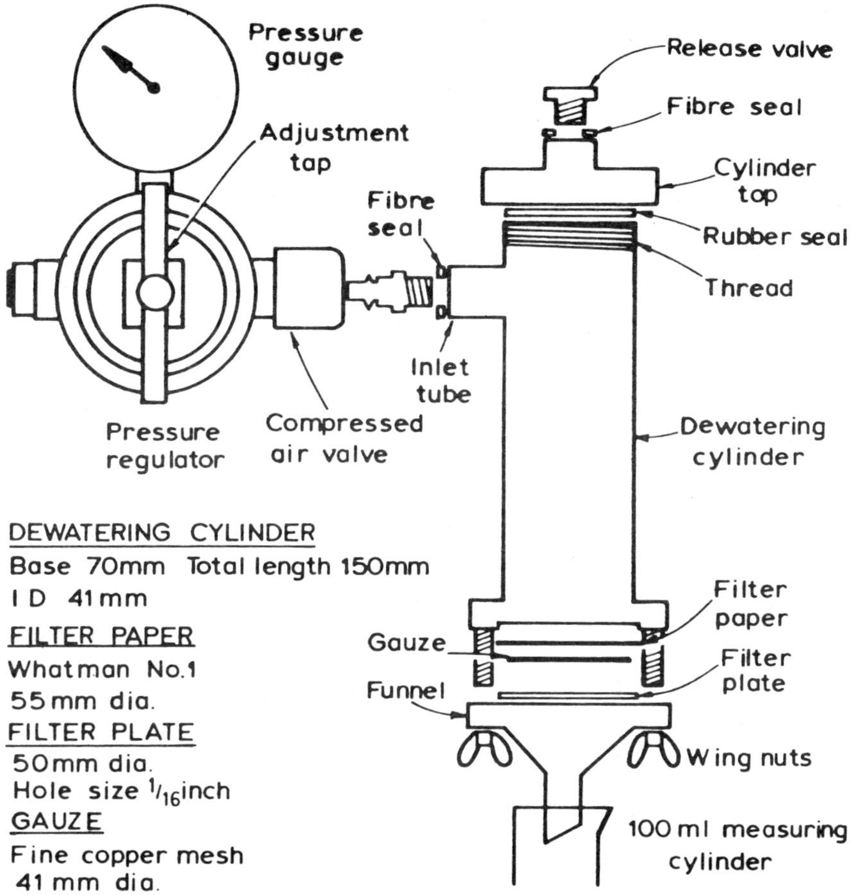

Fig. 6.2 Specific resistance apparatus suitable for high pressure work

Calculation

An MC of 96% ≅ 40 kg/m^3 for the solids concentration C. 675 mm Hg = 133.3 x 675 Pa or 90 kN/m^2 (kPa). Filtration area = $\pi(0.05)^2/4 = 1.96 \times 10^{-3}$m^2. Assuming the filtrate to be water at 20 °C, $\mu = 1 \times 10^{-3}$ Ns/m^2. Values of θ/V against V give the plot shown in Fig. 6.3. Usually a few points close to the origin have to be ignored in order to obtain a straight line. In this case $b = 1.88 \times 10^{12}$ s/m^6. Therefore

$$r = \frac{2bPA^2}{\mu C} = \frac{2 \times 1.88 \times 10^{12} \times 90 \times 10^3 \times (1.96 \times 10^{-3})^2}{10^{-3} \times 40}$$

$$= 3.3 \times 10^{13} \text{ m/kg}$$

□ □ □

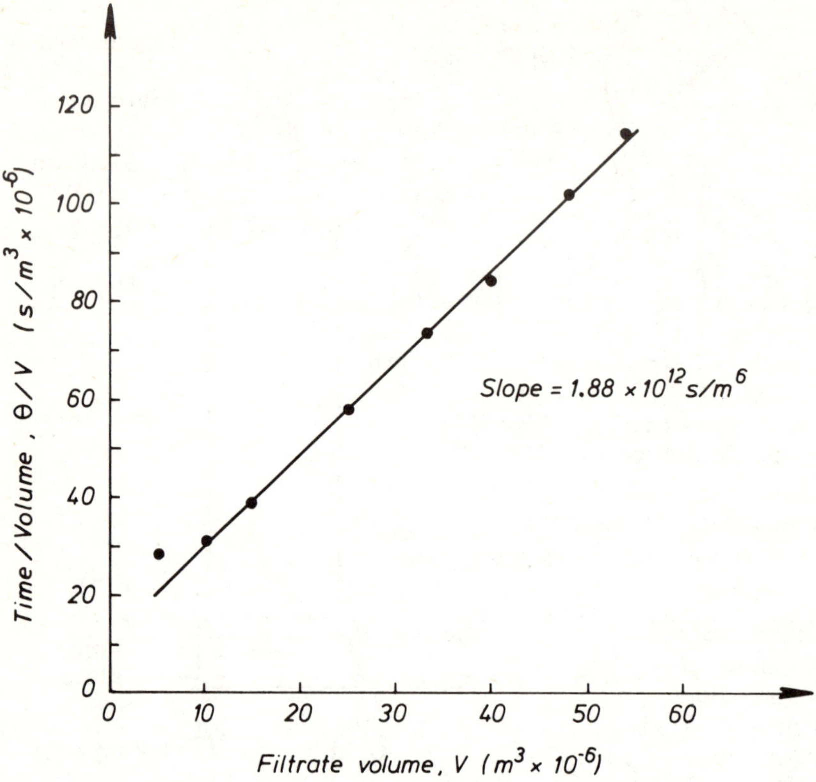

Fig. 6.3 θ/V versus V diagram for determination of specific resistance

6.3.1 Specific resistance units

Prior to metrication, pressures were often read off gauges as g/cm^2 and thus r was quoted in units of s^2/g, e.g.

$$r = \frac{2bPA^2}{\mu C} = \frac{\text{s}}{(\text{cm}^3)^2}\,\frac{\text{g}}{\text{cm}^2}(\text{cm}^2)^2\,\frac{\text{cm s}}{\text{g}}\,\frac{\text{cm}^3}{\text{g}} = \text{s}^2/\text{g} \tag{6.10}$$

In fact the gauges indicated grammes force/cm^2; thus 1 g force/cm^2 read on the gauge was 981 dynes/cm^2. To change a quoted figure of r of 10^9 s^2/g to cm/g requires a multiplier of 981 cm/s^2 and to convert the value of 9.81 x 10^{11} cm/g to m/kg requires a further factor of 10. Thus 'old' values of r quoted in s^2/g require to be multiplied by 9810 to convert them to m/kg.

6.3.2 Sludge cake compressibility

The value of r for most wastewater sludges changes with pressure according to the relationship

$$r = r_0 P^S \tag{6.11}$$

where r_0 is a constant and S is a coefficient of compressibility, which usually varies from 0.5 to 1.1. Knowledge of S allows r to be corrected for any pressure or allows comparison of 'r' values determined for different sludges at different pressures.

Example 6.10
Determinations of r at different pressures give a slope of 0.65 for a plot of log r versus log P for sludge A. The value of r at 90 kN/m^2 is 1.5 x 10^{13} m/kg for sludge A. A value quoted in the literature is 2.3 x 10^9 s^2/g at 2000 g/cm^2 for sludge B. Compare the dewaterability of the two sludges.

Calculation
Literature value (B): a pressure of 2000 g/cm^2 is in effect 2000 x 981 dynes/cm^2 or 196 kN/m^2 and an r value of 2.3 x 10^9 s^2/g is 9810 x 2.3 x 10^9 m/kg or 2.3 x 10^{13} m/kg.

Determined value (A): a value for r of 1.5 x 10^{13} m/kg at 90 kN/m^2 is equivalent, at 196 kN, to

$$\frac{r_{196}}{1.5 \times 10^{13}} = \frac{r_0}{r_0}\left(\frac{196}{90}\right)^{0.65}$$

or r_{196} = 2.5 x 10^{13} m/kg. The specific resistance of sludge B quoted in the literature is therefore lower than an equivalent value for sludge A.

□ □ □

It should be noted that a standard pressure of 49 kN/m^2 (500 gf/cm^2) has often been used when quoting values of r in the literature.

6.4 Sludge dewatering

Three common methods of sludge dewatering are used; vacuum filtration, pressure filtration and centrifugation. All operate on treated wastewater sludges in which the specific resistance has been reduced by a factor of 50 or more.

6.4.1 Vacuum filtration

Gale [4] has suggested a formula relating yield from a vacuum filter (shown diagrammatically in Fig. 6.4) and specific resistance, i.e.

$$Y = \left(\frac{2PC_l A_f}{\mu r \theta_R}\right)^{1/2} f_c \tag{6.12}$$

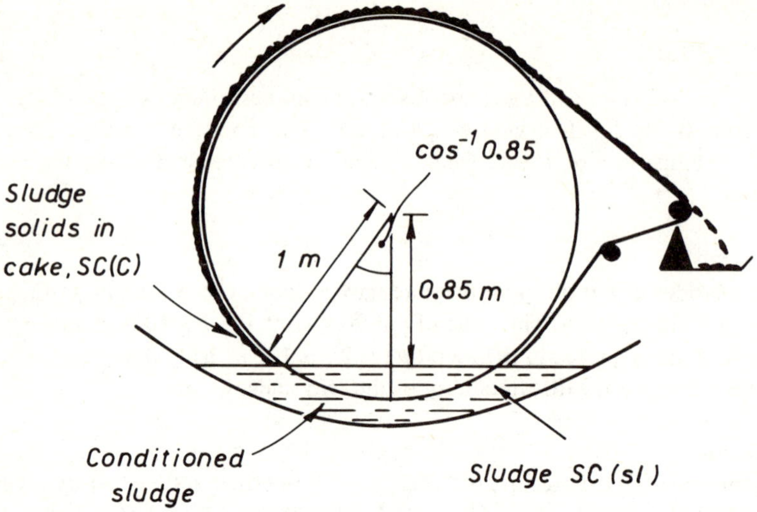

Fig. 6.4 Schematic diagram of vacuum filter

where Y is the yield of dry SS (kg/m^{2}s); A_f is the fraction of filter area used in cake formation, i.e. that area below the sludge surface (m^2); θ_R is the time per revolution (s); C_l is the mass of solids/unit volume of *liquid* in the sludge (kg/m^3); f_c is the (mass of liquid/unit mass of sludge)/(mass of filtrate obtained/unit mass of sludge dewatered) or

$$f_c = \frac{1 - \mathrm{SC(Sl)}}{1 - \mathrm{SC(Sl)/SC(C)}} \tag{6.13}$$

where SC(Sl) is the solids content of the sludge (kg/kg) and SC(C) is the solids content of the cake before drying begins, i.e. as the cake leaves the sludge bath.

Example 6.11

A vacuum filter is 2 m in diameter with cycle time of 150 s/revolution. The initial sludge solids content is 3% and scrapings of cake emerging from the sludge bath show 10% SC. The filter dips 0.15 m into the sludge which is at 20 °C ($\mu = 10^{-3}$ Ns/m^2). A pressure difference of 90 kN/m^2 is developed across the filter using a conditioned sludge with specific resistance 5 x 10^{11} m/kg at 90 kN/m^2. Calculate the filter yield.

Calculation

$$f_c = \frac{1 - \mathrm{SC(Sl)}}{1 - [\mathrm{SC(Sl)/SC(C)}]} = \frac{1 - 0.03}{1 - 0.03/0.10} = 1.39$$

$$C_l = \left(\frac{0.03}{1 - 0.03}\right) 10^3 = 30.9 \text{ kg/m}^3$$

From Fig. 6.4

$$A_f = \frac{\cos^{-1}(0.85)}{180} = 0.18$$

Therefore

$$Y = \left(\frac{2 \times 90 \times 10^3 \times 30.9 \times 0.18}{10^{-3} \times 5 \times 10^{11} \times 150}\right)^{1/2} 1.39$$

$$= 5.1 \times 10^{-3} \text{ kg/m}^2\text{s}$$

□ □ □

The effect of using an unconditioned sludge with $r = 5 \times 10^{13}$ is to reduce the yield to 5.1×10^{-4} kg/m^{2}s, all other values being held constant.

6.4.2 Pressure filtration

One empirical equation used in the determinaton of filter press drying of wastewater sludges is

$$T = k\frac{r\,[\mathrm{SC(F)} - \mathrm{SC(Sl)}]^2}{P\,.\,\mathrm{SC(Sl)}\,[1 - \mathrm{SC(Sl)}]} \tag{6.14}$$

where T is the time to press to a specified final cake solids content SC(F), and k is a constant of proportionality.

T is usually measured in hours. Values of k vary considerably, however, for the calculation below, k is taken as 1.3×10^{-6}.

Example 6.2

(*a*) For a conditioned sludge of 3% SC calculate the pressing times to obtain cakes of 20, 30 and 40% SC given that $P = 550$ kN/m^2 and the specific resistance of the conditioned sludge, measured at 550 kN/m^2, is 5×10^{12} m/kg. Assume $k = 1.3 \times 10^{-6}$.

(*b*) Calculate the same values for input sludge solids of 5% SC.

Calculation

(*a*) For SC(F) = 0.20

$$T = 1.3 \times 10^{-6} \times \frac{5 \times 10^{12}}{550 \times 10^3} \times \frac{(0.20 - 0.03)^2}{0.03(1 - 0.03)}$$

$$= 11.7 \text{ h}$$

For final cake solids of 30% SC and 40% SC the times are 29.6 and 55.6 h, respectively.

(*b*) If the input sludge solids increases to 5% solids after conditioning, the time to press to 20% solids is given by

$$T = 1.3 \times 10^{-6} \times \frac{5 \times 10^{12}}{550 \times 10^{3}} \frac{(0.20 - 0.05)^2}{0.05(1 - 0.05)}$$

$$= 5.6 \text{ h}$$

and times for final cake solids content SC(F) of 30 and 40% for the same input would be 15.6 and 30.5 h, respectively.

□□□

Variations in specific resistance values operate in a proportional manner; thus reducing r by a factor of 100 by chemical treatment achieves 100 times reduction in T. However, variations in P are not so straightforward, since the effect of sludge compressibility becomes important.

Example 6.13

A sludge treated with ferric chloride gives a specific resistance value at 49 kN/m^2 of 10^{12} m/kg ($S = 0.6$) and when treated with cationic polyelectrolytes gives the same specific resistance at 49 kN/m^2 but with compressibility coefficient of $S = 1.0$. Calculate the ratio of the pressing times for the two sludges at a pressure of 700 kN/m^2.

Calculation

If $S = 0.6$ the specific resistance at 700 kN/m^2 will be

$$r_{700} = 10^{12} \times \left(\frac{700}{49}\right)^{0.6} = 4.9 \times 10^{12} \text{ m/kg}$$

for $S = 1.0$

$$r_{700} = 10^{12} \left(\frac{700}{49}\right)^{1.0} = 1.4 \times 10^{13} \text{ m/kg}$$

Therefore, the ratio of the pressing times is given by

$$\frac{T_{(S=0.6)}}{T_{(S=1.0)}} = \frac{4.9 \times 10^{12}}{1.4 \times 10^{13}} = 0.35$$

□□□

It is therefore clear that treatment of sludges with polyelectrolytes ($S = 0.9$–1.1) gives too high a coefficient of compressibility to be used in high pressure filter presses. High pressure processes almost always use inorganic Al or Fe salts together with lime, as sludge conditioners ($S = 0.4$–0.6).

6.5 Incineration

Incineration of sludges, possibly mixed with domestic refuse, is often used where the water content can be lowered to a point where the combustion becomes autothermic, i.e. self-sustaining. In order to achieve efficient combustion it is necessary to add air of the stoichiometric requirement, this is referred to as excess air. For autothermic combustion the heat produced from the calorific value of the solids in the sludge less any heat losses must be sufficient to raise the moisture in the sludge cake to the exit temperature. The actual percentage of the heat available to do this depends upon the wall and ash losses and the heat lost in the gases emitted from the heat exchanger (if fitted.) Detailed calculations are given elsewhere [5, 6], however, the principles can be illustrated using assumed values for the heat available.

6.5.1 Stoichiometric oxygen requirements

Theoretical values for the oxygen required for combustion can be calculated from analytical data. Excess air is then added based on experience for a particular incinerator.

Example 6.14

Given the formula $C_5H_7NO_2$ for waste activated sludge, calculate the volume and mass of air required for combustion assuming 100% excess air is needed.

Calculation

For the above formula we have:

Element	Number of atoms	Atomic weight		Total element mass/g molecular weight
Carbon	5	12	=	60
Hydrogen	7	1	=	7
Nitrogen	1	14	=	14
Oxygen	2	16	=	32
				113

or an analytical make up of 53.1% C; 6.2% H; 12.4% N and 28.3% O. Therefore, in 1 kg of combustible material (CM) we have approximately 0.53 kg C which reacts thus:

$$C + O_2 \rightarrow CO_2 \tag{6.15}$$

$$12\text{ g} + 32\text{ g} \rightarrow 44\text{ g}$$

or

$$0.531\text{ kg C} + \frac{32}{12} \times 0.531\text{ kg } O_2$$

i.e. 1.42 kg O_2 are required to burn the carbon.

Similarly for hydrogen

$$H_2 + O \rightarrow H_2O \tag{6.16}$$

$$2\ \text{g} + 16\ \text{g} \rightarrow 18\ \text{g}$$

$$0.062 + 0.50$$

The 0.12 kg N requires no oxygen since it is emitted as N_2 gas. Therefore the oxygen requirement is

$$1.42\ \text{kg} + 0.50\ \text{kg} - 0.28\ \text{kg (from sludge)} = 1.64\ \text{kg oxygen}$$

At 100% excess air, the oxygen in the excess air is 1.64 kg. At NTP, 22.4 l oxygen contains 32 g O_2, therefore 1.64 kg is equivalent to 1.15 m^3 O_2. In air this is equivalent to 1.15/0.21 or 5.5 m^3 air.

Air mass calculation

0.21 x 22.41 oxygen contains 0.21 x 32 g O_2 and 0.79 x 22.4 l nitrogen contains 0.79 x 28 g N_2. Thus 22.41 air contains 6.72 g O_2 and 22.12 g N_2. Thus in normal air, 1 kg O_2 is associated with 22.12/6.72 kg N_2 or (22.12 + 6.72)/6.72 kg air; i.e. 1 kg O_2 is equivalent to 3.29 kg N_2 or 4.29 kg air. Thus 1.64 kg O_2 = 7.04 kg air.

Therefore, in total, 3.28 kg O_2 are required for input in the form of 11.0 m^3 air/kg combustible material or 14.1 kg air/kg CM.

□□□

6.5.2 Heat budget

Knowing the specific heats of the N_2, air, CO_2 etc. in the exit gas, and estimating the wall, ash and other heat losses enables a heat budget calculation to be carried out [7]. In the example below, an estimate of the gross heat required for water evaporation is made in order to simplify the procedure.

The heat H required to raise 1 kg water from an inlet temperature T_I to an exit temperature T_E where $T_E > 100\ ^\circ\text{C}$ is approximately

$$H = 4.18(100 - T_I) + 2257 + 2.09(T_E - 100)\ \text{kJ} \tag{6.17}$$

where the 2257 kJ is the latent heat of vaporization of water.

If 1 kg of combustible material has a calorific value of K kJ/kg CM, and Z is the fraction of heat from this calorific value which is lost at walls, in ash and exit gases etc., the mass of water which may be raised to the exit temperature is $(1 - Z)(K/H)$ kg H_2O per kg CM incinerated. This can be used to calculate whether a sludge of given SC will burn without extra fuel.

Example 6.15
A mixed sludge has a calorific value of 22 000 kJ/kg CM, is 80% organic and has been dewatered to 75% MC. Calculate whether autothermic combustion is possible:

(a) assuming 40% heat loss with a heat exchanger producing an exit temperature of 450 °C

(b) assuming 60% heat loss with no heat exchanger and an exit temperature of 800 °C.

Calculation
The heat required to raise 1 kg water to an exit temperature of 450 °C from, say, 15 °C, is

$$H_{450} = 4.18(85) + 2257 + 2.09(350) = 3340 \text{ kJ}$$

and similarly to 800 °C (H_{800}) is 4080 kJ.

In case (a) 0.6 x 22 000 or 13 200 kJ/kg CM is available after losses, i.e. there is sufficient heat to raise 13 200/3340 or 3.95 kg H_2O to 450 °C. Therefore if 1 kg CM (or 1 kg organics) is associated with 3.95 kg H_2O, the sludge will just continue to support combustion. Now, 1 kg organics is associated with 0.2/0.8 kg or 0.25 kg inorganics for a sludge which is 80% organic material. Thus the composition of sludge which will just support autothermic combustion is

$$\frac{1 + 0.25}{1.25 + 3.95} = 0.24(24\% \text{ SC})$$

i.e. the 75% MC sludge satisfies the conditions.

For condition (b) the heat available is 8800 kJ and this is equivalent to 2.16 kg H_2O. Thus the cake composition would have to be

$$\frac{1.25}{1.25 + 2.16} = 0.37(37\% \text{ SC})$$

Autothermic combustion cannot be sustained with a sludge of 75% MC and in this case the temperature would gradually drop (unless extra fuel were added) until combustion ceased.

□ □ □

6.6 Return liquor streams

In many forms of sludge treatment a highly polluting liquor is produced which is pumped back to the works inlet in some cases exerting a significant BOD or SS load. Detailed calculations of loads for a wide range of sludge treatment processes have been presented elsewhere[8], the examples below are intended to illustrate the basic methods of such calculations. Table 6.4 gives values of BOD and SS,

Table 6.4 Average BOD and SS concentrations in liquor streams from sludge treatment processes

Process	BOD concentration (mg/l)	SS concentration (mg/l)
Filter press (raw sludge)	2000	200
Filter press (conditioned raw sludge)	1800	150
Filter press (heat conditioned sludge)	6000	200
Filter press (digested sludge)	1400	100
Drying beds (digested sludge)	100	150

Vacuum filtration liquors have similar concentrations to those for filter presses, however the final SC is < 25%.

which, although showing wide variations in practice, may be used for illustration. In the following examples the per capita daily production of BOD and SS are taken as 0.060 kg and 0.090 kg, respectively.

Example 6.16
A conditioned sludge is pressed from 6% to 40% SC producing a liquor containing 1800 mg/l BOD. Calculate the increased inlet BOD load and secondary BOD load owing to return of the liquid stream.

Calculation
The liquor change per kg dry sludge solids is given by

$$\left(\frac{100}{6} - 1\right) - \left(\frac{100}{40} - 1\right) = 14.2 \text{ kg liquor}$$

or for 0.09 kg solids/day, 1.28 kg, i.e. 1.3 l/cap . day.

At 1800 mg/l BOD this represents $1800 \times 1.3 \times 10^{-6}$ or 0.0023 kg BOD/cap . day or approximately 3.8% of the inlet BOD. However, since most would pass through the primary settlement tank, the load on the secondary processes, assuming 40% removal of the inlet BOD, would be increased by $[0.0023/(0.6 \times 0.06)] \times 100\%$ or 6.4%.

□ □ □

Heat treatment liquors can be particularly troublesome.

Example 6.17
The sludge in this example is heat treated at 5% SC prior to filter pressing to 45% SC. Assuming the filtrate to contain 6000 mg BOD/l, calculate the increased BOD load to the secondary processes.

Calculation
The overall liquid change for 0.09 kg solids input is

$$0.09\left[\left(\frac{100}{5}-1\right)-\left(\frac{100}{45}-1\right)\right]\ \text{kg}$$

or approximately 1.6 l/cap . day.

The BOD load is $1.6 \times 6000 \times 10^{-6}$ kg = 0.0096 kg BOD/cap . day, which, compared with 0.036 kg/cap . day BOD in the flow after primary settlement represents an increased loading of 27%. In this case, should the liquor be returned too rapidly to the works inlet, it could overload the secondary plant.

□ □ □

6.7 Conditioning of sludges

Sludges are usually conditioned prior to vacuum filtration, centrifugation or filter pressing. Since sludges vary in source, type, composition and alkalinity, the majority of processes are based upon local, empirical data. One common conditioning process is the use of chemicals, which are usually added as a percentage of the dry sludge solids. Common chemicals include aluminium chlorohydrate $Al_2(OH)_4Cl_2$, aluminium and ferric chlorides $AlCl_3$, $FeCl_3$, chlorinated copperas $FeCl_3 + Fe_2(SO_4)_3$ mixture and lime and copperas. An example of the calculation of dosage rates is given below.

Example 6.18
A sludge of 4% SC is to be conditioned using a lime and copperas mixture, the lime slurry and the copperas solution contain 10% lime and 25% copperas w/v, respectively. Calculate the volumes of slurry and solution required to dose 10 m^3 sludge at 15% of the sludge dry solids content. The ratio of lime to copperas is 5 : 3 as dry weights.

Calculation
At 4% SC 10 m^3 sludge contains approximately 10 x 40 kg or 400 kg dry solids. At 15% of dry solids, 0.15 x 400 kg or 60 kg total lime and copperas solids are required, i.e. (5/8) x 60 kg or 37.5 kg lime and (3/8) x 60 kg or 22.5 kg copperas. A lime slurry of 10% w/v contains 100 kg/m^3 and hence, 37.5/100 m^3 or 375 l of lime slurry is required together with 22.5/250 m^3 or 90 l of copperas solution.

□ □ □

Because $FeCl_3$ is not commonly available in commercial quantities it is sometimes necessary to oxidize copperas to the ferric state by using chlorine to produce a mixture of $FeCl_3$ and $Fe_2(SO_4)_3$ according to the expression

$$6FeSO_4 \,.\, 7H_2O + 3Cl_2 = 2FeCl_3 + 2Fe_2(SO_4)_3 + 42H_2O \qquad (6.18)$$

The minimum amount of Cl_2 necessary can be calculated from the above expression.

Example 6.19

Calculate the stoichiometric (minimum) amount of Cl_2 necessary to treat 500 l of 25% copperas solution.

Calculation

The molecular weights of $FeSO_4 \cdot 7H_2O$ and Cl_2 are

$$\begin{array}{cccc} \text{Fe} & \text{S} & O_4 & 7 \times H_2O \end{array}$$

$$(55.9 + 32.1 + 64.0) + (7 \times 18)$$

or 278.0 for $FeSO_4 \cdot 7H_2O$ and

$$2 \times \text{Cl} \quad \text{or } 71.0 \text{ for } Cl_2 \text{ gas}$$

$$(2 \times 35.5)$$

Thus 278 g $FeSO_4 \cdot 7H_2O$ requires 71.0 g Cl_2 x (3/6) or 35.5 g Cl_2 . 500 l of a 25% solution of copperas contains 500 x 0.25 kg or 125 kg copperas and would require (125/278) x 35.5, i.e. 16.0 kg Cl_2.

□□□

Questions

Question 6.1

The following results were obtained from a specific resistance test:

Time (s)	132	295	511	792	1122
Filtrate (ml)	3.0	5.0	7.0	9.0	11.0

The pressure difference $= 97.5$ kN/m^2
Sludge moisture content $= 92\%$
The absolute viscosity of the filtrate $= 10^{-3}$ Ns/m^2
The filtration area $= 4 \times 10^{-4}$ m^2

Calculate the specific resistance to dewaterability using Equation (6.8).

Question 6.2

Sewage from a population of 50 000 has a yearly mean temperature of 15 °C. The sludge is produced at a rate of 0.09 kg dry SS/head day at 98% moisture content, and is anaerobically digested at 32 °C for 25 days. Suggestions have been made (*a*) to reduce the retention time to 20 days, or (*b*) to provide a consolidation tank to reduce the inlet sludge moisture content to 95% MC. Given the information below, evaluate individually, in terms of heat energy, the probable effects of adopting each suggestion.

Solids digestion = 30% for 25 days or 26% for 20 days
Gas production = 1 m^3/kg solids digested
Gas calorific value = 22 x 10^3 kJ/m^3
Boiler efficiency = 70%
Sludge specific heat = 4.18 kJ/kg °C
Equivalent heat loss of digester = 0.6 °C/day

Question 6.3

A sewage sludge of 75% volatile content has a calorific value (heat of combustion) of 23 250 kJ/kg combustible (organic) material. It is incinerated and gives 58% of the calorific value as heat available to raise the cake moisture from the inlet temperature (15 °C) to the gas exit temperature of 430 °C. Calculate the solids content of the cake necessary to support autothermic combustion, given 3200 kJ/kg is required to raise water at 15 °C to steam at 430 °C.

Question 6.4

A conditioned sludge of 2% solids content gave a value for *r* of 5 x 10^{11} m/kg at 90 kN/m^2 when the absolute viscosity of the filtrate was 1.3 x 10^{-3} Ns/m^2. A vacuum filter 2 m diameter operating at 90 kN/m^2 is to be used to dewater the sludge. The sludge TWL is 0.8 m below the filter axis. Scrapings gave a value of 15% for SC(C). Calculate the filter yield if the cycle time is 200 s.

References

1* Ames, Crosta, Babcock Ltd. Publication 100, *Simplex sludge digestion plants,* Heywood, UK.
2 Swanwick, J. D. and Davidson, M. F. (1961), *Water and Waste Treat.*, 8, 386.
3 Coackley, P. (1958), in *Biological Treatment of Sewage and Industrial Wastes* (eds. J. McCabe and W. W. Eckenfelder, Jun.), van Nostrand–Rheinhold, New York.
4 Gale, R. S. (1971), *J. Inst. Wat. Pollution Contr.*, **70**, 114–18.
5* Barnes, D. and Wilson, F. (1978), *Chemistry and Unit Operations in Sewage Treatment,* Applied Science Publishers, London, Chapter 9.
6 Gale, R. S. (1972), in *Proceedings of the Symposium on Incineration of Refuse and Sludge* (ed. P. Helliwell), Southampton University Press.
7 Archer, G. A. (1972), in *Proceedings of the Symposium on Incineration of Refuse and Sludge* (ed. P. Helliwell), Southampton University Press.
8 Clough, G. F. C. (1977), *J.Inst. Wat. Pollution Contr.*, **76**, 452–58.

* Suggested further reading.

7
Pumping

Nomenclature

Symbol	*Definition and units*	*Dimension*
h_s	Static lift, resistance at zero flow, m	L
h_{pipe}	Pumping head losses owing to pipe friction, m	L
Q	Liquid flow rate, m^3/h unless stated	L^3T^{-1}
n	Mannings coefficient, $s/m^{1/3}$	$TL^{-1/3}$
d	Pipe diameter (internal), mm or m as stated	L
l	Pipe length, km unless stated	L
h_{ftg}	Pumping head losses due to bends, valves etc., m	L
C	Friction loss coefficient, dimensionless	—
V	Velocity of flow in pipe, m/s	LT^{-1}
g	Gravitational constant, m/s^2	LT^{-2}
H	Total pumping head (m), $H = h_s + h_{pipe} + h_{ftg}$	L
P	Pump input power, kW	ML^2T^{-3}
Q_{gas}	Gas flow rate to air lift pump, m^3 free air/s	L^3T^{-1}
C'	Constant, m^3/s	L^3T^{-1}
m	Constant	
s	Depth of submergence, m	L
L	Lift height (air input to top of rising main), m	L

7.1 Pump requirement

One relatively straightforward method of choosing a pump for a job is to consider a simple pumping network; to calculate the head losses and hence to determine the duty range on the pump characteristic curve. Considering the pumping system in Fig. 7.1, the pump has to overcome both static lift (h_s), i.e. initially from A (TWL in the sump) to B (BWL in the reservoir) or just prior to when pumping ceases, from C (BWL in sump) to D (TWL in reservoir), and also the frictional resistance caused by the moving liquid. This latter term can be expressed as a head loss (m) thus

$$h_{pipe} = l\,7952 \times 10^9 \times \left(\frac{Q \times n}{d^{8/3}}\right)^2 \qquad (7.1)$$

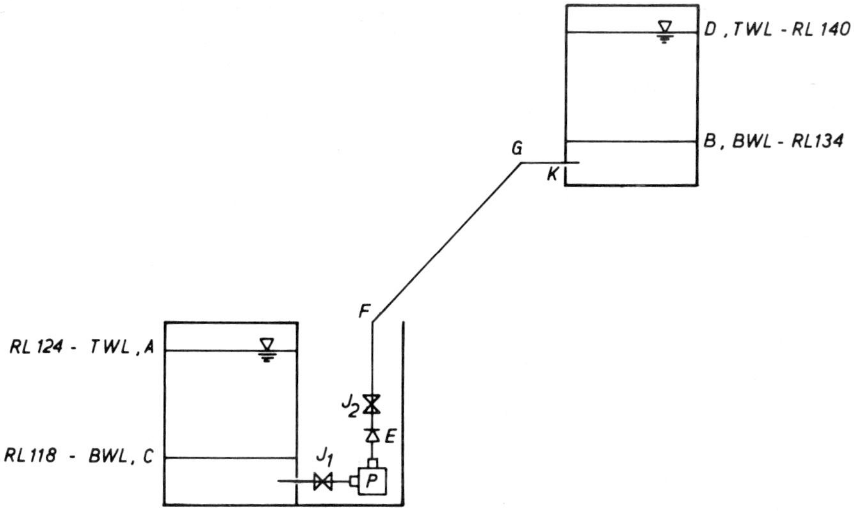

Fig. 7.1 Simplified pumping network

where h_{pipe} is the head loss (m) in a pipe of length l (km); Q is the flow rate (m^3/h), n is the Mannings coefficient (usually between 0.011 to 0.013 for iron pipes – see Table 7.1) and d is the pipe diameter (mm).

Where a pipe has bends, T junction etc., a further series of losses occur which are evaluated in the form

$$h_{fittings} = \frac{CV^2}{2g} \tag{7.2}$$

where C is a coefficient which can be estimated from tables of the type shown in Table 7.2. The following example shows how the losses may be estimated in the case of water, sewage or sewage effluents.

Table 7.1 Friction coefficients for use with Manning's formula

Pipe material	Condition	Value of n
Coated iron and steel pipes	new	0.010
Uncoated iron pipes	new	0.011
Galvanized iron pipes	new	0.012
Vitrified clay pipes	after some weeks of service	0.012
Iron and steel pipes	after use	0.013
All pipes	with imperfect joints and in bad condition	0.015

[Courtesy of R. E. Bartlett[1] and Applied Science Publishers]

Table 7.2 Velocity head loss coefficients in fittings and valves

Fitting	Coefficient (C)
Belmouth	0.10
Bend	0.25 to 0.50
T-junction	1.00
Sluice valve	0.15
Reflux valve (single flap)	0.70 to 2.00*

* The value of C for reflux valves increases as the velocity decreases.
[Courtesy of R. E. Bartlett[1] and Applied Science Publishers]

Example 7.1

Water is to be pumped by a rotodynamic (centrifugal) pump at a rate of 100 m^3/h from the sump in Fig. 7.1 to the reservoir. Assume bends at F, and G ($C = 0.30$), a reflux valve at E ($C = 1.0$), two fully open sluice valves J_1 and J_2 ($C = 0.15$) and a leaving loss ($C = 1$) at K. Calculate the head losses and find the extreme duty points on the pump characteristic curve (usually supplied by the pump manufacturer). Assume a rising main diameter of 150 mm and a Manning coefficient of 0.012 with pipe length of 250 m.

Calculation

Static lift: minimum = 134.0 – 124.0 = 10.0 m

maximum = 140.0 – 118.0 = 22.0 m

Friction head losses in the pipe can be calculated from Equations (7.1) and (7.2). At 25 m^3/h Equation (7.1) gives

$$h_{pipe} = 0.25 \times 7952 \times 10^9 \left(\frac{25 \times 0.012}{150^{2.67}}\right)^2 = 0.4 \text{ m}$$

Table 7.3 shows similar values for other flow rates.

Table 7.3 Pipe friction losses

(i)	(ii)	(iii)	(iv)
Q	$0.25 \times 7952 \times 10^9 \left(\frac{Qn}{d^{8/3}}\right)^2$	For h_s minimum	For h_s maximum
m^3/h	m	(ii + 10.0)m	(ii + 22.0)m
0	0	10.0	22.0
25	0.4	10.4	22.4
50	1.7	11.7	23.7
75	3.9	13.9	25.9
100	6.9	16.9	28.9
125	10.7	20.7	30.7
150	15.4	25.4	37.4

In order to assess the friction loss due to fittings, the velocity must first be calculated from

$$V(\text{m/s}) = \frac{Q\,4 \times 10^6}{3600\ \pi d^2} \qquad (7.3)$$

e.g. at 25 m^3/h

$$V = \frac{25}{3600}\ \frac{4 \times 10^6}{\pi(150)^2} = 0.39 \text{ m/s}$$

Therefore the head loss due to fittings is

$$h_{ftg} = \frac{V^2}{2g}(C_1 + C_2 + C_3 + C_4 \ldots)$$

where $C_1, C_2, C_3 \ldots$ are coefficients from Table 7.3.
Thus

$$h_{ftg} = \frac{0.39^2}{2 \times 9.81}\,[(2 \times 0.3) + 1.0 + (2 \times 0.15) + 1.0]$$

$$= 0.02 \text{ m}$$

At other flow rates the head losses due to fittings are as shown in Table 7.4 together with the total head loss for the system $H = h_s + h_{ftg} + h_{pipe}$.

□□□

Table 7.4 Head losses due to fittings and total losses

(i)	(ii)	(iii)	(iv)	
Q	$V = \frac{4Q \times 10^6}{3600\pi\,(150)^2}$	h_{ftg}	$h_{ftg} + h_{pipe} + h_s$	
			minimum	maximum
m^3/h	m/s		m	m
0	0	0	10.0	22.0
25	0.39	0.02	10.4	22.4
50	0.79	0.09	11.8	23.8
75	1.18	0.21	14.1	26.1
100	1.57	0.36	17.3	29.3
125	1.96	0.57	21.3	33.3
150	2.36	0.82	26.2	38.2

An imaginary pump characteristic curve for a rotodynamic pump is given as curve X–X in Fig. 7.2. The pump would be chosen from a manufacturers catalogue after specifying the maximum pumping head and maximum flow. The

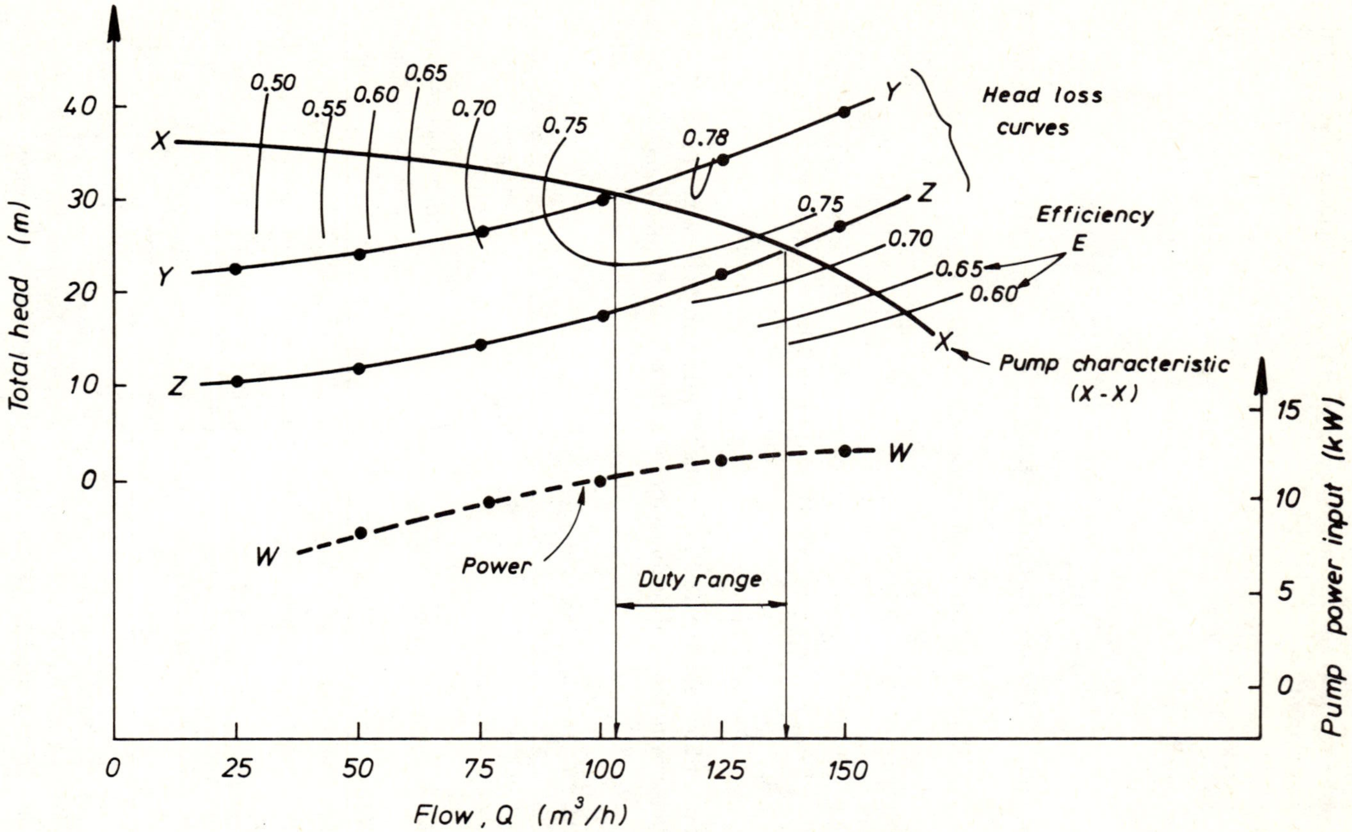

Fig. 7.2 Rotodynamic pump characteristics

curve in Fig. 7.2 actually represents one size of impeller. In practice the impeller can be machined down to give better match between a particular pump and the system requirements.

The system resistance versus flow curves $Y-Y$ and $Z-Z$ have been superimposed on the pump characteristic curve in Fig. 7.2. Perpendiculars dropped from the minimum and maximum duty points onto the abcissa give the expected minimum and maximum flow rates.

7.2 Pump input power

Efficiency contours have been superimposed on the pump characteristic curve in Fig. 7.2. These can be used to give the required input power to the pump if this is not supplied by the pump manufacturer. If Q m^3/s is raised H(m) against gravity at an efficiency (E) then (assuming 1 m^3 H_2O = 1000 kg) the power (P) expended is

$$P = \frac{QHg}{E} \tag{7.4}$$

the units of P are 10^3 J/s, i.e. kilowatts.

Example 7.2
From the pump characteristic curve given in Fig. 7.2 construct a table of pump input power requirements for given flow rates, efficiencies and pumping heads.

Calculation
From the characteristic curve, a flow rate of 50 m^3/h (0.014 m^3/s) can be achieved at 34.5 m head at an efficiency of, say, 0.58. The power input necessary is therefore

$$P = \frac{0.014 \times 34.5}{0.58} \times 9.81 = 8.2 \text{ kW}$$

The result of similar calculations are given in Table 7.5 and plotted as on Fig. 7.2 as pump input power requirements curve $W-W$.

Table 7.5 Input power requirements

Q (m^3/h)	$Q/3600$ (m^3/s)	H (m)	E	P (kW)
50	0.014	34.5	0.58	8.2
75	0.021	33.0	0.71	9.6
100	0.028	30.5	0.77	10.9
125	0.035	26.5	0.76	12.0
150	0.042	20.5	0.68	12.4

□ □ □

7.3 Pumping of viscous sludges

Water and dilute suspensions (< 99.5% MC) are Newtonian fluids, i.e. a plot of shear stress versus shear rate gives a straight line through the origin whose slope is the single value of the absolute viscosity. Unfortunately, sludges do not follow this relationship and often show a decreasing apparent viscosity with increasing shear rate (the apparent viscosity being the ratio of shear stress to shear rate for a plot which is not a straight line through the origin). Furthermore, the sludge changes its characteristics with changes in moisture content. Data on particular sludge characteristics are therefore important and it is stressed that the approach given below involves some risk. However, in the absence of such information, Bartlett[1] has suggested that the frictional head losses for different MC sludges may be estimated by multiplying the values calculated for water by a suitable factor. Bartlett's suggested relationship of friction factor with MC is shown in Fig. 7.3. This curve can be used to calculate the head losses when pumping sludges.

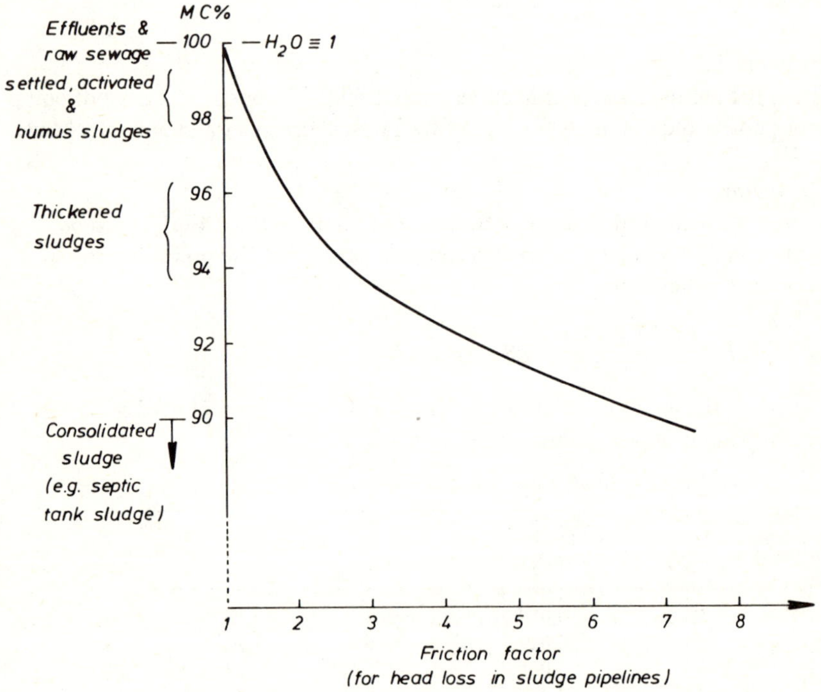

Fig. 7.3 Sludge pumping friction factors (by courtesy of R. E. Bartlett and Applied Science Publishers [1])

Example 7.3
Using the same conditions as in Example 7.1 with the exception that a sludge of MC 94% has to be pumped, calculate the head losses in the system. No change occurs in the static lifts which remain at 10.0 m (minimum) and 22.0 m (maximum). However, both the pipe friction losses and the fittings friction losses have to be multiplied by the factor 2.7 (as Fig. 7.3).

Calculation
The new minimum head loss at 100 m^3/s flow is now $H = 10.0$ m + (2.7 x 7.3) m = 29.7 m. Similar calculations for different flows are summarized in Table 7.6.

Table 7.6 Flow rate versus head losses for 94% MC sludge

Flow rate (m^3/h)	Total head loss minimum (m)	maximum (m)
0	10.0	22.0
25	11.1	23.1
50	14.9	26.9
75	21.1	33.1
100	29.7	41.7
125	40.5	52.5
150	53.7	65.7

The pump shown in Fig. 7.2 is inadequate and a more powerful pump must be chosen.

□□□

7.4 General

In practice the rotodynamic pump may alter speed as the load changes. In this case the actual pump characteristic may be a curve parallel to that in Fig. 7.2, e.g. higher if the speed increases, thus giving a greater flowrate for a given pumping head. The changes in speed depend upon the prime mover, e.g. diesel engine, electric motor etc. The overall efficiency also depends on the efficiency of the pump and the prime mover. Thus if the pump requires 10.8 kW input power to deliver 100 m^3/h and is driven by a prime mover of 84% efficiency, the overall requirement is 12.9 kW.

Positive displacement pumps are occasionally used in wastewater treatment; their characteristic shows little reduction in flow with increasing pressure. The calculations are similar to those previously carried out for rotodynamic pumps, but in practice they are thought to supply a specified flow rate independent of frictional losses.

7.5 Air lift pumps

These are commonly found on small scale extended aeration plants as a means of returning settled activated sludge to the aeration tank. Larger capacity pumps are sometimes used for difficult situations where the low efficiency is offset by the lack of moving parts. For larger pumps the reader is referred elsewhere [1], the discussion below refers to a typical small installation.

Many parameters influence the performance characteristics but an approximate relationship between liquid pumping rate Q (m^3/s) and volumetric rate of free gas input Q_{gas} (m^3/s) in the region of maximum efficiency is of the type [2]

$$Q = m \log_{10} Q_{gas} + C' \qquad (7.5)$$

where C' and m are constants which vary with rising main diameter and with ratios of S/L where S is the depth from air input to TWL and L is the height from air input to top of riser as shown in Fig. 7.4.

Values of m and C' are given for different pipe sizes and S/L ratios in Table 7.7. It should be noted that these data were obtained from a pump of length $L = 4.25$ m, and therefore should not be applied to pump lengths much less than

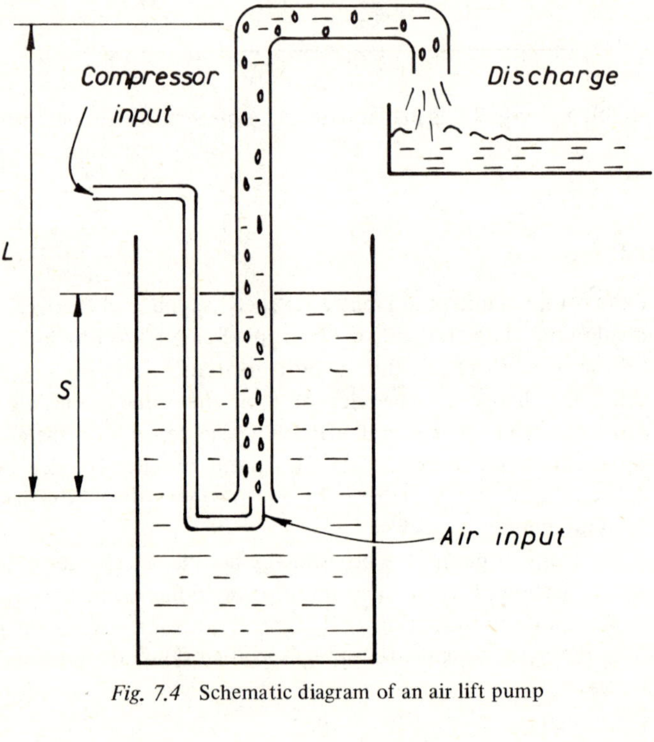

Fig. 7.4 Schematic diagram of an air lift pump

Table 7.7 Coefficients for air lift pumps (for $L = 4.25$ m)

S/L ratio	pipe diameter (mm)	m –	C' –
0.71	25.4	0.0003	0.0013
	38.1	0.0015	0.0050
	50.8	0.0029	0.0087
0.53	25.4	0.0002	0.0010
	38.1	0.0010	0.0033
	50.8	0.0019	0.0053

3.5 m or much greater than 5 m. The values in Table 7.7 can be used to estimate the air requirement.

Example 7.4
Calculate the air requirement for an air lift system pumping 0.7 l/s water in a 50.8 mm pipe. The vertical rising main is 3.3 m tall and the submergence is 2.34 m, i.e. S/L is 0.71.

Calculation
For this situation m is 0.0029 and C' is 0.0087. Therefore

$$\frac{0.7 \times 10^{-3} - 0.0087}{0.0029} = \log_{10} Q_{gas}$$

therefore $Q_{gas} = 1.7 \times 10^{-3}$ m^3/s or 1.7 l/s free air.

□ □ □

In practice the pump may be returning settled activated sludge of over 1% SC and the resistance will be considerably greater than that of water. In addition, circumstances may arise where a greater liquid flow rate is required. Both these points argue for a compressor which is larger than the theoretical value by a factor of 2 to 2.5 and higher still for an intermittent duty pump.

It should be noted that the larger the values of S/L and pipe diameter, the more efficient is the pump. Note also that the relationship in Equation (7.5) holds for air flow rates up to approximately 5.0 l/s for the 50.8 mm pump or up to 2.0 l/s for 25.4 mm.

Questions

Question 7.1
The pump whose characteristics are given in Fig. 7.2 is to be used to pump water through a 100 m pipe of diameter 0.150 m. There are two 45° bends ($C = 0.40$),

two open valves ($C = 0.15$) and a tank entrant loss ($C = 1.1$). The static lift is constant at 20 m. Calculate the discharge rate and the pump input power if $n = 0.012$.

Question 7.2

A small air lift pump is used to return settled activated sludge through a 50 mm diameter pipe at 0.5 l/s. The rising main dips 2.5 m into the settlement tank and rises to 1 m above the tank TWL. Calculate the air requirement if the sludge requires twice the air rate for water.

References

1* Bartlett, R. E. (1974), *Pumping stations for Water and Sewage*, Applied Science Publishers, London.

2 Sharma, N. D. and Sachdeva, M. M. (1976), An air-lift pump study. *J. Instn. Engrs. (India)*, **56**, 61–3.

* Suggested further reading.

8

Small sewage treatment plants

Nomenclature

Symbol	*Definition and units*	*Dimension*
C	Capacity of tank, l	L^3
N	Population number	–
W	Width of tank, m	L
L	Length of tank	L
V	Volume of media, m^3	L^3
Q_H	Flow in a pipe at fractional depth of flow H, l/s, m^3/s, as stated	L^3T^{-1}
Q_F	Flow in a pipe when full, l/s, m^3/s as stated	L^3T^{-1}
H	Fractional depth of flow in pipe	–
n	Manning's coefficient of roughness, $s/m^{1/3}$	$TL^{-1/3}$
V_H	Velocity of flow in pipe at fractional depth of flow H, m/s	LT^{-1}
V_F	Velocity of flow in pipe when full, m/s	LT^{-1}
F	BOD loading, kg/day	MT^{-1}
M	Sludge solids in aeration tank, kg	M
θ_c	Sludge age, days	T
W_s	Waste sludge production rate, kg/day	MT^{-1}
$\dot{P}$	Power required to compress air, W	ML^2T^{-3}
$\dot{m}$	Rate of transfer of compressed air, kg/s	MT^{-1}
C_p	Specific heat of air at constant pressure, J/kg K	$L^2T^{-2}K^{-1}$
ΔT	Temperature rise in compression process, K	K
p_I	Air pressure at inlet to compressor, N/m^2	$ML^{-1}T^{-2}$
p_O	Air pressure at compressor outlet, N/m^2	$ML^{-1}T^{-2}$
T_I	Air temperature at compressor inlet, K	K
T_O	Air temperature at compressor outlet, K	K
γ	Ratio of specific heats of air at constant pressure and constant volume	–

8.1 Introduction

Increasingly stringent effluent standards are being applied to those discharges from small communities, campsites, hotels etc. which previously had suffered only dilution in the receiving water, or some simple form of 'treatment' such as septic tanks. This chapter is intended to give some simple design procedures for the various treatment methods applied to small communities. Since a more comprehensive text is available [1] the design examples will be kept short and discussion limited. The format of theoretical statement followed by worked example is somewhat inappropriate for this chapter since most operational criteria are empirically based; these criteria will be given as the design process unfolds.

8.2 Septic tank systems

Septic tanks are simply long retention time tanks where solids deposit out in an anaerobic bottom layer; a certain amount of grease and other solids forms a crust on the surface while the liquid intermixes with the products of decomposition from the sludge layer.

In the UK the septic tank is not accepted as a treatment method by itself unless large effluent dilutions are possible, since it cannot meet the usual Royal Commission standards (20 mg/l BOD; 30 mg/l SS). In this case, therefore, the tank is usually followed by a percolating filter. In other countries, the septic tank is a means of removing gross solids and thus lengthens the life of the following land infiltration trenches. The two approaches are sufficiently different to be considered separately.

8.2.1 Septic tank – percolating filter (UK)

The tank is to be a two compartment system serving a single household designed in accordance with CP 302 [2]. The minimum population is 4. The capacity

$$C = (180N + 2000)\ \mathrm{l} \tag{8.1}$$

where N is the population. Thus, in this case $C = 2.72\ \mathrm{m}^3$.

The tank is split into tanks of capacity $(2C/3)$ and $1C/3$, i.e. $1.81\ \mathrm{m}^3$ and $0.91\ \mathrm{m}^3$. The tanks would have a (common) flat floor and be 1.5 m deep below TWL, the first tank would be rectangular in plan with the width $W = 1/2$ (length L). Thus the two dimensions are:

1st tank

$$W = \sqrt{\left(\frac{1.81}{2 \times 1.5}\right)} = 0.78\ \mathrm{m};\ L = 2W = 1.56\ \mathrm{m}$$

at 1.5 m liquid depth with, say, 0.7 m headspace.

2nd tank

$W = 0.78$ m; $L = 0.78$ m

the depth and headspace would be as above.

Details of interconnection between tanks, inlet and outlet pipes, covers, etc. are given in CP 302[2].

(*a*) *Percolating filter*

CP 302 suggests a volume of media V of 1 m^3 per head for $N < 10$, i.e. 4 m^3 in this case. Other suggestions[3] are $V = 1.5N^{0.8}$ or 4.6 m^3 for 4 persons. Assuming 4 m^3 media and a depth of 1.8 m, the surface area is 2.2 m^2, e.g. a bed 1.1 m x 2.1 m is sufficient, loaded by notched channels fed from a two-way tipper which fills and tips to one side and then fills and tips to the other side of the bed.

(*b*) *General*

(i) A 100 mm pipe laid to the 'old' rule of 1 in 40 (0.025) would give a full flow velocity of 1.1 m/s and a flow of 8.9 l/s for a Manning's coefficient 'n' of 0.012. A daily peak flow of say 1.3 l/s could be relied upon from WC flushing, which, from Fig. 1.3, would give Q_H/Q_F of 0.15; thus $H = 0.25$; $V_H/V_F = 0.72$, see Chapter 1, Section 1.3, and the velocity of flow at 1.3 l/s would be 0.72 x 1.1 or 0.79 m/s. This pipe and slope would be satisfactory to be laid from the house to the tank.

(ii) Variations in effluent quality may incline the water authority to insist on tertiary treatment, in which case, two plots of grassland at a slope of 1 in 80 to 1 in 100 of dimensions 3.5 m square (half to be in use at any time) would probably be satisfactory (see also Section 5.2.4).

(iii) The tank should be emptied at yearly intervals, leaving some sludge as an innocculum for anaerobic digestion.

(iv) The cost of individual septic tanks is such that investigations into the possibility of community sewage treatment should be pursued prior to installation.

8.2.2 Septic tank – soakaway

In many countries the septic tank–soakaway system is common. The tank itself is often a single unit although since carryover of solids is more important than in the previous case, a better design would be to adopt a two tank unit as before. The actual design procedures for the tank may be based on number of bedrooms, or on a per capita water usage, however the designs usually accept a minimum retention time of 24 h at average daily flow plus 2000 l capacity to allow for sludge storage etc.

The percolation of the local soil is usually determined by a standard test and the results expressed as min/mm reduction in water level. The surface area of the trench is taken as bottom area and is determined from a curve of the type given in Fig. 8.1, e.g. a population of 4 produces 200 l/cap . day of septic tank effluent, the percolation rate is determined on the local soil (moderately free draining clayey loam) at 16 min/mm. The bottom area loading required for this soil is 30 l/m^2day. For a square section trench 0.45 x 0.45 m the length required is (200 x 4)/(30 x 0.45) m or approximately 60 m.

'Resting' of a section of the trench has been found to be useful in reducing surface clogging (presumably by aerobic degradation of slimes etc.) and therefore the above trench may be increased to three sections each of 30 m length, any

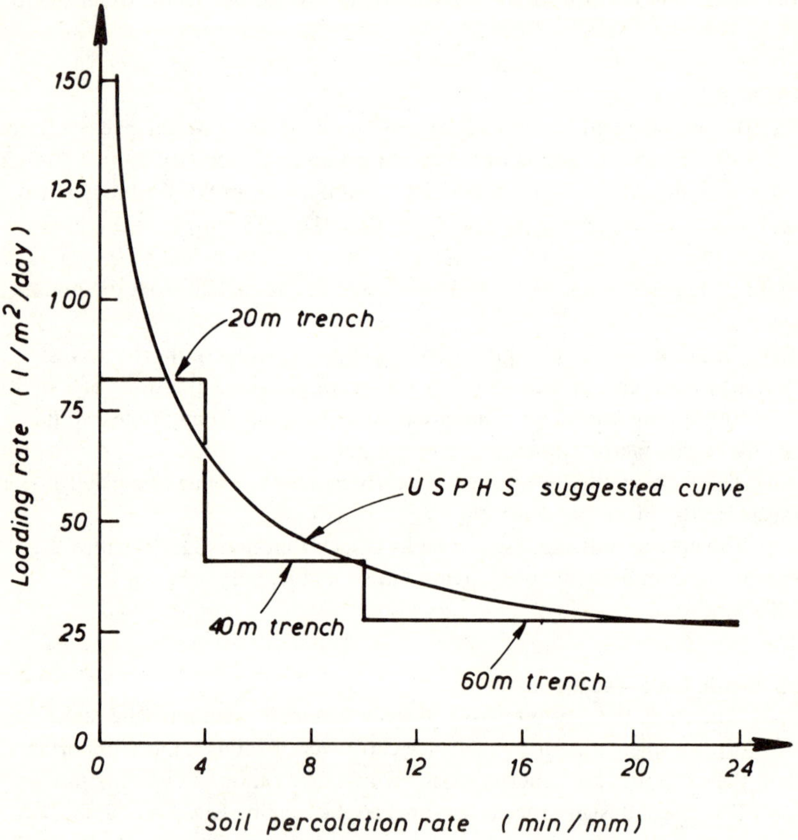

Fig. 8.1 United States Public Health Service and present New Zealand practice soakaway requirements (by courtesy of New Zealand Standards Association)

two being used at one time. The trenches are usually packed with 50 mm diameter gravel around tile drains to a depth of 200–250 mm with topsoil above (mounded to allow rain water run-off.)

8.3 Extended aeration

A common type of scheme which is used for small communities is the 'package' extended aeration (EA) plant. This is usually prefabricated in a factory from steel and transported to the site and commissioned by the distributor. Because of the difficulties of labour and sludge disposal inherent in the use of primary settlement tanks, the package usually omits all preliminary and primary treatment apart from occasional use of a holding sump (to even out flow surges) and the more common use of screens (sometimes omitted or placed after the aeration stage.) The scheme therefore can be represented by the simple schematic diagram shown in Fig. 8.2, which includes an aerobic sludge storage digester stage.

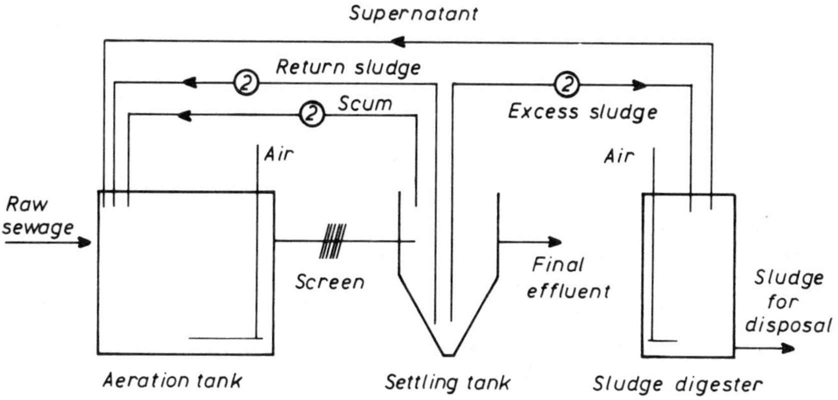

Fig. 8.2 Schematic diagram of small activated sludge plant (by courtesy of A. Johnson Construction Co. Ltd, Glasgow)

The proposed scheme will be based on the flow from 50 houses using UK values for flow, BOD and SS inputs. An aerobic digester will be incorporated into the scheme, to act as a sludge storage facility. Values for the aeration tank MLSS and settlement tank thickened sludge solids are assumed to be 3000 mg/l and 12 000 mg/l, respectively. The effluent criteria required by the River Authority is 15 mg/l BOD and 15 mg/l SS which implies the use of tertiary treatment.

8.3.1 Influent

Population is based on 3.7 persons per house = 185 persons. A water usage per capita, of 200 l/cap . day gives 37 m^3/day.

The BOD contribution per capita at 0.06 kg/cap . day = 11.1 kg/day
The SS contribution per capita at 0.085 kg/cap . day = 15.7 kg/day

The BOD and SS concentrations averaged over the day are 300 mg/l BOD and 424 mg/l SS (the surface water and ground water are assumed to be completely excluded).

8.3.2 Aeration tank

Extended aeration schemes require a retention time of approximately 1.3 days[2]. The aeration volume is therefore 37 x 1.3 m^3 = 48 m^3. To improve aeration efficiency the aerators will be placed 3 m below TWL. The breadth would be 2 m for this tank, giving a rectangular box of approximately 2 x 7.8 m x 3.1 m deep (dimensions are those below TWL)

A common *F/M* ratio range, i.e. BOD (kg/day)/MLSS (kg) is 0.05–0.1/day. The daily BOD input is 11.1 kg and assuming the MLSS is 3000 mg/l the *F/M* ratio is 11.1/(48 x 3000 x 10^{-3}) kg/day = 0.08 kg/day which is satisfactory.

The usual calculation for sludge waste estimates 0.5 kg MLSS per kg BOD removed, i.e. 0.5(300–20) x 10^{-3} x 37 kg/day or 5.18 kg/day assuming 20 mg/l BOD in the effluent stream. The sludge age θ_c = (48 x 3)/5.18 = 28 days (it is common not to include the sludge present in the settlement tank in these calculations).

8.3.3 Oxygen and air requirement

Assuming an effluent BOD concentration prior to the tertiary treatment stage of 20 mg/l, the overall removal of BOD would be (300 – 20) x 37 x 10^{-3} kg BOD/day or 10.4 kg BOD/day.

Assuming 2.0 kg O_2/kg BOD, this would be equivalent to 20.8 kg O_2 required, which, at say 4% transfer efficiency, would be 520 kg O_2/day. Assuming 24 l of air at 1 atm pressure and 15 °C contains 32 x 0.21 g O_2, the air requirement would be

$$\frac{520}{32 \times 0.21 \times 10^{-3}} \times 24 \times 10^{-3}\ \mathrm{m^3/day} = 1860\ \mathrm{m^3/day}$$

see Chapter 4, Section 4.4.

An alternative estimate could be based on the CP 302 requirement of 10 m^3 air per capita at 3 m depth, i.e. 1850 m^3/day. A figure of 1850 m^3/day will be assumed for compressor rating for aeration alone.

8.3.4 Settlement tank

The settlement tank may be designed on the basis of overflow rates (22 m^3/m^2 day)[2], i.e. 1.7 m^2 surface area, and a retention time of 4.5 h, i.e. 7 m^3

capacity. The weir loading should be $< 100\ m^3/m$ day thus requiring a weir length of 0.4 m.

However, several points of design are important and have an effect on the above values.

(*a*) The overflow rate includes the effect of surface skimmers and eductors. Assuming a skimmer operating intermittently at a rate of 7 l/min the surface area will need to be increased by this amount, e.g. 22 m^3/m^2 day loading would require 37/22 or 1.7 m^2 settlement tank area; 7 l/min over 1.7 m^2 is equivalent to 0.07 mm/s or 6 m^3/m^2 day. The corrected loading rate is therefore 16 m^3/m^2 day giving a design area of 37/16 m^2 or 2.3 m^2.

(*b*) If the settlement tank is formed as a trapezoidal (in side section) addition to the aeration tank, the breadth would be 2 m and the surface length 1.2 m. A weir length up to 2 m is therefore available.

(*c*) A check should be made that the solids loading kg MLSS/m^2 day is < 100. In this case the loading rate is

$$\frac{37 \times 10^3 \times 3000 \times 10^{-6}}{2.3} = 48\ \text{kg/m}^2\text{day}$$

which is satisfactory.

8.3.5 Activated sludge return pump

A positive return pump system employing a 50 mm rising main airlift pump is probably suitable. The total volume to be returned is (37 x 3000)/12 000 m^3/day. i.e. 9.25 m^3 neglecting the daily waste flow of sludge (0.11 l/s). The total air requirement based on Example 7.4 in Chapter 7 would be

$$\text{antilog}_{10}\left(\frac{0.11 \times 10^{-3} - 0.0087}{0.0029}\right)$$

$$= 1.1 \times 10^{-3}\ \text{m}^3\text{/s or } 1.1\ \text{l/s}$$

Allowing for a factor of 2.5 to overcome sludge resistance the airlift pumps would require 2.8 l/s air or 237 m^3/day.

8.3.6 Aerobic digester (see Chapter 4, Section 4.3.3)

The per capita daily excess sludge production is 0.028 kg from before. Assuming that this is to be stored over 50 days at 10 000 mg/l solids concentration, the total sludge volume, allowing for 30% digestion of solids, would be

$$\frac{0.028 \times 10^6 \times 50 \times 0.7}{10\,000} = 98\ \text{l/cap . day}$$

CP 302 suggests 90 l/cap . day for a small scheme although other countries will accept lower volumes. However, at 90 l/cap . day the total volume is 16.7 m^3.

The air supply to the digester is often estimated as 30 m^3/m^3day or 3.3 m^3/cap . day which gives total aeration requirements of 501 m^3/day and 611 m^3/day, respectively. A value of 500 m^3/day will be assumed to be satisfactory.

8.3.7 Compressor requirements

The total compressor requirements are

	m^3/day
aeration tank	1850
airlift pumps	237
digester	500
	2587 m^3/day or approximately 2600 m^3/day

The compressor motor size can be estimated from thermodynamic calculations of the type given below:

The power P (W) required to compress a gas can be shown to be approximately given by

$$P = \dot{m} C_p \Delta T \qquad (8.2)$$

for an ideal gas where $\dot{m}$ is the mass flow rate of gas transferred (kg/s); C_p is the specific heat of air at constant pressure (approximately 1005 J/kg K) and ΔT is the rise in temperature (K).

For a compressor transferring air at atmospheric pressure to atmospheric pressure + 3 m water (sewage) depth, the ratio of outlet pressure p_O to inlet pressure p_I is approximately

$$\frac{p_O}{p_I} = \frac{(101\,000 + 3000 \times 9.81)}{101\,000} = 1.3 \qquad (8.3)$$

The ratio of outlet temperature T_O to inlet temperature T_I for an ideal (isentropic) compression process is given by

$$\frac{T_O}{T_I} = \frac{p_O}{p_I}^{(\gamma - 1/\gamma)} \qquad (8.4)$$

where γ is the ratio of specific heats at constant pressure and at constant volume, e.g. 1.4 for air. Thus

$$\frac{T_O}{T_I} = 1.078 \qquad (8.5)$$

The temperature rise, assuming an inlet temperature of 288 K, is 0.078 x 288 K or 22.5 K.

The volume of air transferred is 2600 m^3/day or 0.030 m^3/s, which, at a density of 1.2 kg/m^3, is equivalent to 0.036 kg/s. Therefore the ideal rate of work (power) required by the compressor is

$$P = mC_p\Delta T = 0.036 \times 1005 \times 22.5 \text{ W}$$

or 814 W

Assuming 50% efficiency, the energy requirement is approximately 1630 W. A 2 kW drive is therefore the minimum suitable size motor.

8.3.8 Tertiary treatment (see Chapter 5, Section 5.2.2)

The simplest scheme, providing land is available, is to operate grass plots with distribution and collection channels on a slope of 1 in 100. An area of 1.5 m^2/cap. in use at any time gives 280 m^2 with a further 280 m^2 duplicate 'resting area'. Dimensions of 17 m x 34 m long would be suitable. Removals of 50% of both SS and BOD can be assumed. Further details of extended aeration plant designs are given in a more specialist publication[1].

8.4 Rotating discs

The rotating disc process incorporates a large number of lightweight discs made of braced expanded polystyrene, supported plastic netting or expanded metal mesh. The discs rotate in the settled sewage, alternately contacting air and sewage. The aerobic oxidation proceeds in a similar manner to a percolating filter. The discs are grouped into stages with baffles being used to prevent short circuiting between stages. For small scale units, the disc unit is preceded by an Imhoff tank in which settlement of the raw sewage occurs, the settled sludge being cold digested and stored in a separate chamber. Rising gas is prevented from gaining access to the settlement compartment. The disc unit is followed by a humus settlement tank, the settled sludge being pumped to the Imhoff tank at regular (e.g. daily) intervals. Fig. 8.3 shows a typical small scheme (see Chapter 4, Section 4.6 for a design of a large scale rotating disc unit).

8.4.1 Small scheme

The process will be based on a small community of 200 persons discharging 200 l/cap . day with 0.06 kg BOD/cap . day and 0.085 kg SS/cap . day.

8.4.2 Imhoff tank

The digestion zone in this tank has to store both the primary sludge (70% of the inflow SS) and also the returned secondary sludge (0.8 kg SS/kg BOD removed in the disc stage). Desludging is usually carried out four times per year and an assumption is made that the sludge will have an average MC of 93%.

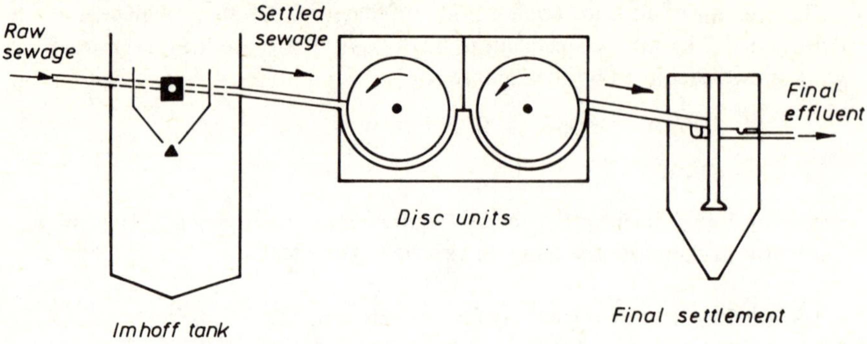

Fig. 8.3 CJB Type W rotating disc plant (by courtesy of CJB developments Ltd, Portsmouth)

If 33% of the BOD is removed in the primary tank, the sludge produced per day is:

$$\text{primary sludge} = 0.7 \times 0.085 \times 200 = 11.9 \text{ kg SS}$$

$$\text{secondary sludge} = 0.8 \times 0.04 \times 200 = 6.4 \text{ kg SS}$$

i.e. a total of 18.3 kg SS/day.

If this is stored for 100 days, the sludge mass will be 1830 kg. At 70 kg/m^3 (93% MC) this is equivalent to 1830/70 m^3 or 26 m^3. The settlement zone may be based on a retention time of 4.5 h at DWF, i.e.

$$\frac{200 \times 200}{10^3} \times \frac{4.5}{24} \text{ m}^3 \text{ or } 7.5 \text{ m}^3$$

A total tank volume of 34 m^3 is suitable. If a circular concrete tank 3.5 m deep and 4 m in diameter is used, this gives adequate capacity. A suspended steel settlement tank 3.5 m long by 1.5 m wide and 1.2 m deep could be fitted into the concrete cylinder. The side walls below 1.2 m would slope downwards at 60° to the horizontal with a slot to allow solids to pass to the digestion chamber. A deflector would be fitted just below the slot to prevent rising gas from entering the settlement tank. The total volume would be > 7.5 m^3. The surface loading rate for the settlement tank would be

$$\frac{200 \times 200}{10^3 \times 3.5 \times 1.5}$$

or 7.6 m^3/m^2day at DWF and 30 m^3/m^2day at 4 DWF. Assuming a 1.5 m width weir the respective weir loading rates would be 27 m^3/m day and 107 m^3/m day. These are satisfactory, see Chapter 3, Table 3.4.

8.4.3 Disc unit

Disc units for small schemes are often loaded at approximately 6.5 g BOD/m^2 disc area calculated using both sides of a disc. The BOD to the disc is 67% of the inlet BOD, i.e.

$$0.06 \times 200 \times 0.67 = 8.04 \text{ kg}$$

Using 3 m diameter discs the number required is

$$\frac{8040 \times 4}{6.5 \times 2\pi 9} = 88 \text{ discs (rounding up)}$$

Using 4 stages, this gives 22 discs per stage (see Chapter 4, Table 4.6). Assuming 40% of the disc area is submerged at any time, the depth of flow would be approximately the submerged depth of the disc, 1.31 m, plus up to 50 mm clearance between the discs and the tank invert. A tank shaped to a radius of 1.36 m would be suitable with baffles inserted after each stage of 22 discs.

8.4.4 Humus tank

Since this tank is not intended to act as a long term storage, a suitable capacity would be 4 h retention plus 1500 l, i.e.

$$200 \times 200 \times \frac{4}{24} + 1500 \text{ l or } 8.2 \text{ m}^3$$

Using a length to breadth ratio of 2:1 and a minimum depth of 1.2 m gives approximate dimensions of 1.85 m x 3.70 m x 1.2 m deep. The surface area is 6.85 m^2 which gives a surface loading rate of 5.9 m^3/m^{2}day at DWF and would be well below the conventional limit of 30 m^3/m^{2}day at 4 DWF. The weir loading rate, assuming a full width weir, would be 22 m^3/m day at DWF and 88 m^3/m day at 4 DWF. Reference to Chapter 3, Table 3.4 shows that these fall well within conventional practice limits.

8.5 Oxidation ditches

Oxidation ditches are extended aeration schemes operating in continuous looped channels 1 to 2 m deep. Aeration is usually achieved by means of a rotor which straddles the channel. The degree of turbulence and oxygenation imparted to the liquid is controlled by the depth of immersion of the rotor blades. A minimum forward velocity of 0.3 m/s and a minimum oxygen input of 2 kg/kg BOD removed are necessary. Several methods of operating the channels are illustrated in Fig. 8.4. The continuous system uses an aeration channel and a separate settlement tank with continuous settled sludge return. The split channel system incorporates the settlement tanks within the ditch while the intermittent system

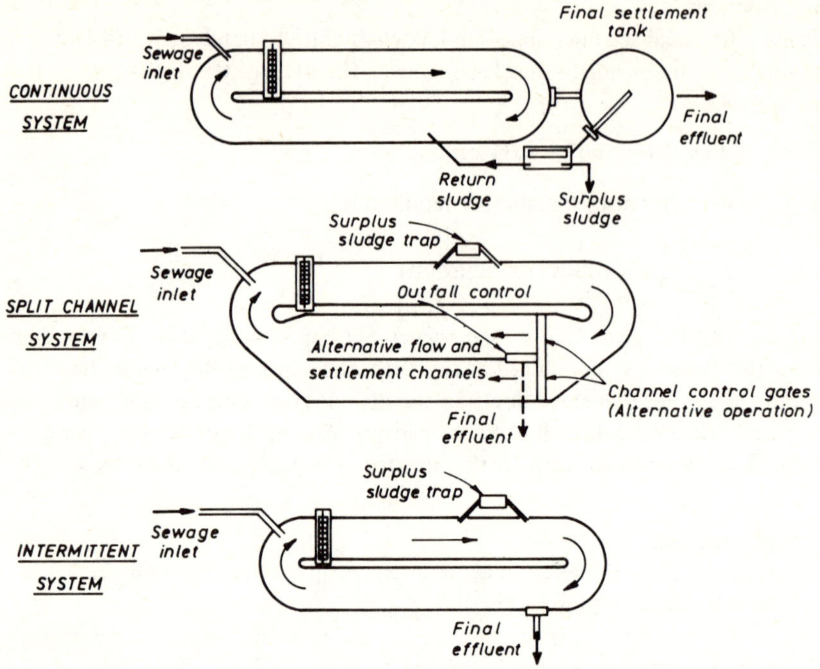

Fig. 8.4 Typical format of Whitehead & Poole oxidation ditches (by courtesy of Whitehead & Poole, Manchester)

uses a cycle of aeration, back rotation to stop the flow, settlement and decantation. This latter system lends itself ideally to a nitrification–denitrification system as discussed in Chapter 5, Section 5.4.

The following calculation is intended to show how the major design criteria can be established for a simple 'racecourse' shaped system using a trapezoidal channel with sides sloping at 45° as in Fig. 8.5.

The per capita daily flow and BOD input are 200 l and 0.055 kg, respectively. The total population served is to be 500 and the plant is to achieve a 20:30:: BOD:SS standard. Dimensions of the rotor usually determine the dimensions of the channel and therefore these latter are assumed to be those quoted for a particular manufacturer's rotor[3]. Details from a brochure produced for this particular rotor type are reproduced in Table 8.1. The MLSS is assumed to be 5000 mg/l.

Calculation

CP 302 suggests a capacity of 260 l/cap., i.e. a total volume of 130 m^3 and a retention time of > 30 h.

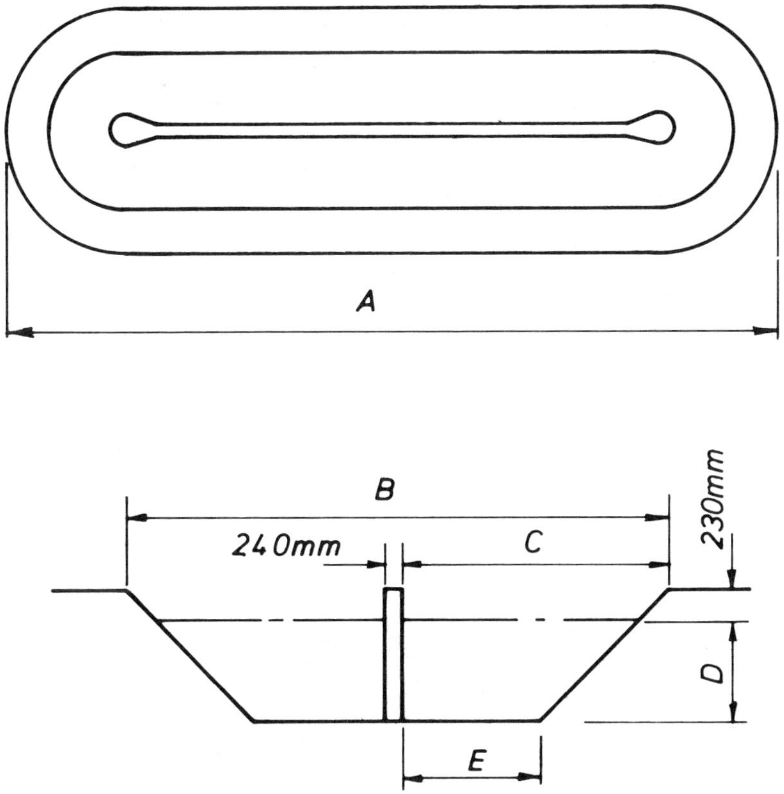

Fig. 8.5 Dimensions referred to in Table 8.1 (by courtesy of Whitehead & Poole, Manchester)

Small tanks such as this are often used with 1 m water depth and a 1/2 channel bottom width of 1.7 m, e.g. see Table 8.1. The volume of the straight part of the 'running track' assuming 45° slope for the sides is $4.4L$ m^3 where L is the length of the straight portion in metres. The volume of the ends of the track can be calculated by considering them to be joined, thus forming the frustrum of a 45° cone of base 2.7 m radius and top 1.7 m radius.

A cone of height h and radius r has a volume $\pi r^2 h/3$. The volume enclosed by the ends of the channels

$$= \frac{\pi}{3}(2.7)^2\ 2.7 - \frac{\pi}{3}(1.7)^2\ 1.7$$

$$= 15.5 \text{ m}^3$$

Table 8.1 The oxidation ditch system utilizing WP–TNO cage rotors [courtesy of Whitehead and Poole [3]]

Load		TNO rotors		Total power (kW)		Aeration tank						
Population Equivalent	BOD (kg/day)	Number and length (m)	Immersion (mm)	Installed	Demand	Number	Total volume (m^3)	A	B	C	D	E
200	11.0	1 × 0.75	70	1.5	0.50	1	52.5	16	5.7	2.73	1.0	1.5
500	27.5	1 × 1.0	118	2.2	0.92	1	131	32	6.1	2.93	1.0	1.7
1000	55.0	1 × 2.0	118	4.1	1.83	1	262	37	8.3	4.03	1.2	2.6
2500	137.5	1 × 4.5	130	7.5	4.55	1	655	44	13.9	6.83	1.4	5.2
5000	275.0	2 × 4.5	130	15	9.17	1	1310	63	15.1	7.43	1.8	5.4
7500	412.5	2 × 7.0	125	22	13.9	1	1965	68	20.1	9.93	1.8	7.9
10 000	550.0	2 × 9.0	130	30	18.4	1	2620	74	24.1	11.93	1.8	9.9
		4 × 4.5	130	30		2		63	15.1	7.43	1.8	5.4
12 500	687.5	2 × 10.5	140	37	22.7	1	3275	82	27.1	13.43	1.8	11.4
		4 × 6.0	122	37		2		64	18.1	8.93	1.8	6.9
15 000	825.0	2 × 14.0	125	45	27.6	1	3930	78	34.1	16.93	1.8	14.9
		4 × 7.0	125	45		2		68	20.1	9.93	1.8	7.9

The channel length required for the straight portion is

$$L = \frac{130 - 15.5}{4.4} = 26.0 \text{ m}$$

and the overall length at TWL is

$$26.0 + 2(2.7) \text{ m or } 31.4 \text{ m}$$

Examples of a particular manufacturer's suggested dimensions are reproduced in Table 8.1.

It should be noted that a criterion suggested prior to CP 302[2] was a loading of $\leqslant$ 210 mg BOD/l ditch capacity per day. Using 0.055 kg BOD/cap . day this gives a volume of $(55 \times 10^3)/210$ or 262 l/cap. However, if the system were to be designed, say, for Australasian conditions, where the BOD could be 0.07 kg/cap . day the volume required would increase to 333 l/cap.

The total BOD loading is 500 x 0.055 kg/day, and, allowing 2 kg O_2/kg BOD per day, the rotor would require an oxygenation capacity (OC) of 55 kg O_2/day. Using a TNO cage rotor of length 1 m (suspended on a gantry across the channel) the OC in kg O_2/mh would be 2.3 kg/mh which would require an immersion of 118 mm (see Fig. 8.6) and require a gross power of 1.2 kW/m. A daily per capita power usage would be 0.058 kWh/cap . day. The *F/M* ratio for the ditch should be at the low end of the range 0.05/day to 0.15/day suggested for

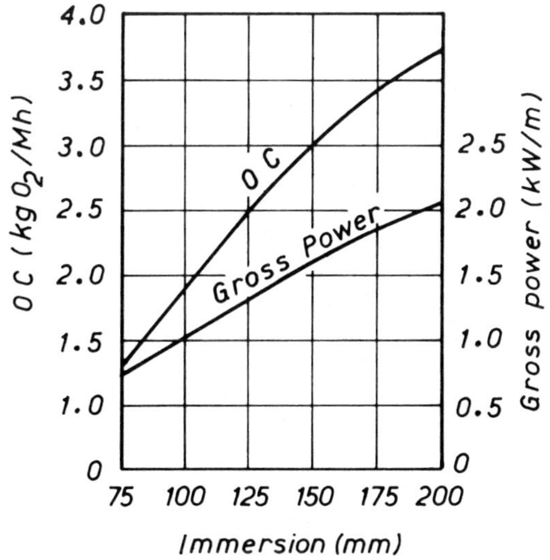

Fig. 8.6 Characteristics of WP TNO cage rotor (by courtesy of Whitehead & Poole, Manchester)

extended aeration schemes. In this case the ratio is

$$\frac{F}{M} = \frac{500 \times 0.055}{5000 \times 10^{-3} \times 130} = 0.042 \text{ /day}$$

This has been found to give satisfactory results in practice. However, if difficulty were found in achieving a sludge with satisfactory settling characteristics, wasting sludge in order to reduce the MLSS to 4000 mg/l would give a F/M ratio of 0.053/day. If 200 l/cap . day of 5000 mg/l MLSS are to be settled, the settlement tank loading would be 300 m^3/day at 3 DWF which would require 13.6 m^2 surface area at 22 m^3/m^2 day. The choice of 3 DWF is somewhat arbitrary but may be considered to be an approximation of the maximum flow rate that will occur for any extended period of time. For further discussion, the reader is referred to a more detailed text [1].

The settled sludge will be returned to the oxidation ditch except for perhaps 30 mg/l MLSS which is discharged over the weir as effluent SS, i.e. 3 kg/day. If the rate of sludge production is 0.5 kg MLSS/kg BOD removed, then assuming 20 mg/l BOD in the effluent (2 kg/day) the sludge production rate W is

$$W = \tfrac{1}{2}(500 \times 0.055 - 2) = 12.8 \text{ kg/day}$$

For a steady state situation, sludge will have to be wasted at a rate of (12.8 – 3) kg/day or 9.8 kg/day.

The sludge age will therefore be

$$\theta_c = \frac{130 \times 5000 \times 10^{-3}}{12.8} = 51 \text{ days}$$

If the sludge is wasted directly as MLSS, the volume to be withdrawn from the ditch would be $9.8/(5000 \times 10^{-6})$ l/day or approximately 2 m^3/day.

References

1* Barnes, D. and Wilson, F. (1976), *Design and Operation of Small Sewage Works*, E. & F. N. Spon, London
2 BS CP 302 (1972), *Small Sewage Treatment Works*, British Standards Institution, London.
3 *The Whitehead and Poole Oxidation Ditch System*, Whitehead and Poole Publications, Manchester.

Further reading

Operation and Management of Small Sewage Works, (1971), HMSO, London.

* Suggested further reading.

9
Receiving waters

Nomenclature

Symbol	*Definition and units*	*Dimension*
L	BOD remaining at time t, mg/l	ML^{-3}
K_1	BOD reaction rate constant to base 'e', /day	T^{-1}
t	Time of flow, day	T
k_1	BOD reaction rate constant to base 10, /day	T^{-1}
k_T	Value of k_1 at T °C, /day	T^{-1}
k_{20}	Value of k_1 at 20 °C, /day	T^{-1}
L_0	BOD ultimate, mg/l	ML^{-3}
T^{L_0}	Value of L_0 at T °C, mg/l	ML^{-3}
20^{L_0}	Value of L_0 at 20 °C, mg/l	ML^{-3}
T	Temperature, °C	°C
C_s	Saturation concentration of oxygen in pure water, mg/l	ML^{-3}
C	Actual oxygen concentration in the stream, mg/l	ML^{-3}
D	Dissolved oxygen deficit $D = C_s - C$, mg/l	ML^{-3}
D_0	Value of D when $t = 0$, mg/l	ML^{-3}
k_2	Reaeration coefficient to base 10, /day	T^{-1}
K_La	Reaeration coefficient to base 'e', /day. Also K_2	T^{-1}
K_2	Reaeration coefficient to base 'e', /day	T^{-1}
L_a	BOD ultimate in Streeter–Phelphs equation when $t = 0$, mg/l	ML^{-3}
D_a	Dissolved oxygen deficit in stream when $t = 0$, mg/l	ML^{-3}
D_c	Dissolved oxygen deficit at critical point, mg/l	ML^{-3}
t_c	Time when critical deficit occurs, day	T
f	Constant, k_2/k_1	–
Δt	Time interval, time of flow between stations, day or h	T

$L_{a(1)}$ and $L_{a(2)}$	Values of L_a at stations (1) and (2), mg/l	ML^{-3}
V	Mean velocity of flow in stream over distance between stations, m/s	LT^{-1}
R	Mean depth in stream between stations, m	L

9.1 Introduction

The effluents from wastewater treatment plants are eventually discharged either into the ground, if the works is very small, a river or lake, or directly to the sea. Disposal to the ground has been dealt with in Chapter 8, Section 8.2.2. The discharge of effluents into marine environments involves not only solution of the complex hydraulic and oceanographic problems of designing and siting an outfall to achieve a specific dilution of the wastewater, but also problems in setting medical and aesthetic standards pertaining to bathing, fishing and commercial shellfish undertakings. The reader is referred elsewhere for further information on this specialist topic[1].

9.1.1 River and lakes

The discharges to rivers and lakes are sometimes specified in terms only of BOD and suspended solids; more stringent specifications may include bacteriological standards, heavy metal concentration limits and specifications as NO_3^-/NH_3 ratios. However, the majority of cases are specified in terms of BOD and suspended solids for low dilutions and suspended solids alone for higher dilutions.

9.2 BOD

The concept of BOD and reaction rates has already been introduced in Chapter 1 section 1.5.1 in which the assumption of a first-order BOD reaction was made, i.e.

$$\frac{dL}{dt} = -K_1 L \qquad (1.39)$$

and

$$L = L_0 \; 10^{-k_1 t} \qquad (1.41)$$

where L and L_0 are the carbonaceous BOD remaining at time t and the ultimate carbonaceous BOD, respectively, and K_1 and k_1 are the reaction rate constants to base 'e' and base 10, respectively.

The change in BOD with temperature is twofold:

(*i*) k_1 changes with temperature as

$$k_T = k_{20} 1.047^{(T-20)} \qquad (1.46)$$

where k_T and k_{20} are values of k_1 at T °C and 20 °C, respectively; and
(*ii*) L_0 also changes:

$$^{T}L_0 = {}^{20}L_0\,[1 + 0.02\,(T - 20)] \tag{9.1}$$

where $^{T}L_0$ and $^{20}L_0$ are the ultimate BOD at T °C and 20 °C, respectively (this change in ultimate load with increasing temperature also appears to occur in anaerobic systems).

The combination of the two effects of temperature changes can cause considerable changes in the total load to receiving waters.

Example 9.1
A BOD_5 (20 °C) test was carried out on a river water sample and gave a BOD_5 value of 20.0 mg/l with a reaction rate constant of 0.18/day. Calculate the BOD_5 at a temperature of 10 °C.

Calculation
At 20 °C, $BOD_5 = L_0\,(1 - 10^{-k_1 5})$

thus

$$L_0(20\,°C) = \frac{20.0}{1 - 10^{-0.9}} = 22.9 \text{ mg/l}$$

and L_0 at 10 °C

$$= 22.9\,[1 + 0.02(-10)]$$

$$= 18.3 \text{ mg/l}$$

k_1 at 10 °C

$$= 0.18 \times 1.047^{(-10)}$$

$$= 0.11 \text{ per day}$$

Thus BOD_5 (10 °C) $= 18.3(1 - 10^{-0.11 \times 5})$

$$= 13.1 \text{ mg/l}$$

□ □ □

9.3 Oxygen saturation and dissolved oxygen deficits

The concentration of oxygen in a stream is of prime importance to the aquatic organisms and hence to the life of the river. Many rivers arise as oxygen saturated, unpolluted, turbulent, fast flowing bodies of water and enter the sea as slow moving, industrially and domestically polluted sewers. In most cases, a crude description of the 'health' of the river can be stated in terms of the 'percentage oxygen saturation', normally found in samples of the water, and the unoxidized BOD.

For fresh water the saturation dissolved oxygen level C_s is given by

$$C_s = 14.61 - 0.394T + 0.007\,714T^2 - 0.000\,0646T^3 \text{ mg/l} \qquad (4.28)$$

where T is the temperature in °C.

The dissolved oxygen deficit (D, mg/l) is the difference between C_s and the actual oxygen concentration in the river (C mg/l), i.e.

$$D = C_s - C \qquad (4.31)$$

From Chapter 4, Section 4.4.1:

$$\frac{\mathrm{d}C}{\mathrm{d}t} = K_L a\,(C_s - C) \qquad (4.30)$$

where $K_L a$ is a constant (day^{-1})

and

$$\frac{\mathrm{d}D}{\mathrm{d}t} = -K_L a\,(C_s - C) \qquad (9.2)$$

This is usually expressed as

$$D = D_0 10^{-k_2 t} \qquad (9.3)$$

where D_0 is the initial dissolved oxygen deficit (mg/l) at time $t = 0$ and k_2 is a constant with units of time^{-1}, (usually day^{-1}); it is usually referred to as a reaeration coefficient and its magnitude greatly influences the rate at which a river can replenish its stock of oxygen after depletion by a polluting discharge.

The value of k_2 increases with temperature thus:

$$T^{k_2} = 20^{k_2} \times 1.0238^{(T-20)} \qquad (9.4)$$

where T^{k_2} and 20^{k_2} represent the values of k_2 at T °C and 20 °C, respectively. Thus, although C_s diminishes with temperature, k_2 increases, hence little difference in rate of change of dissolved oxygen deficit is found in many waters between 0 °C and 30 °C.

Example 9.2
If k_2 for a body of fresh water, initially 0% saturated, is 0.6 per day at 20 °C, calculate the initial rate of change of dissolved oxygen in the water. Calculate also the initial rate of change if the water were originally at 0 °C.

Calculation
From Equation (4.30)

$$\frac{\mathrm{d}C}{\mathrm{d}t} = K_L a\,(C_s - C)$$

and

$$K_L a = 2.3k_2$$

and

$$C_s - C = C_s \text{ when } C = 0$$

Thus, at 0 °C $(C_s - C) = 14.6$ mg/l and at 20 °C $(C_s - C) = 9.3$ mg/l. Also at 20 °C, $k_2 = 0.6$ per day and at 0 °C, $k_2 = 0.6 \times 1.024^{(-20)} = 0.37$ per day.

Thus at 20 °C

$$\frac{dC}{dt} = 2.3 \times 0.6 \times 9.3 = 12.8 \text{ mg/lday}$$

at 0 °C

$$\frac{dC}{dt} = 2.3 \times 0.37 \times 14.6 = 12.4 \text{ mg/lday}$$

The ratio of the two initial aeration rates is 1.03.

□ □ □

9.4 Sag curves in rivers

A very simple river oxygenation model is described by the equation of Streeter and Phelps[2]

$$\frac{dD}{dt} = K_1 L - K_2 D \tag{9.5}$$

i.e. the oxygen deficit will be increased by increasing the polluting load L and diminished by reaeration. Although this model neglects the effects of photosynthetic oxygen production (by algae) and the oxygen depletion by benthic (bottom) deposits, it can be surprisingly accurate, since these two neglected effects often cancel out. Streeter and Phelps also showed that, by integrating Equation 9.5, a general equation for the dissolved oxygen deficit at any time t could be written, namely

$$D = \frac{k_1}{k_2 - k_1} L_a \left(10^{-k_1 t} - 10^{-k_2 t}\right) + D_a\, 10^{-k_2 t} \tag{9.6}$$

where L_a and D_a are the initial ultimate BOD and initial dissolved oxygen deficit in the stretch of the river under consideration and k_1 and k_2 are the oxidation and reaeration rate constants, respectively, taken to base 10. The general form of the equation is a 'sag' curve.

Example 9.3

If k_1 and k_2 for a given river are 0.15 per day and 0.20 per day, respectively, calculate the dissolved oxygen deficits for the flow 6 days downstream if the river is initially 80% saturated and has an ultimate BOD of 10 mg/l. The value of C_s for the river is 12 mg/l.

Calculation

For day 0

$$D_a = 0.2 \times 12 \text{ mg/l, i.e. } 2.4 \text{ mg/l}$$

For day 1

$$D_1 = \frac{0.15}{0.20 - 0.15} 10 (10^{-0.15} - 10^{-0.20}) + 2.4 \times 10^{-0.20}$$

$$= 3.82 \text{ mg/l}$$

For day 2

$$D_2 = 30 (10^{-0.3} - 10^{-0.4}) + 2.4 \times 10^{-0.4}$$

$$= 4.05 \text{ mg/l}$$

Similarly, D_3 = 3.71 mg/l, D_4 = 3.16 mg/l, D_5 = 2.57 mg/l and D_6 = 2.04 mg/l.

□□□

The shape of the curve of C versus time is shown in Fig. 9.1, which clearly indicates a minimum DO concentration somewhere around 1.5 to 2.5 days. This critical value of the dissolved oxygen deficit D_c which is reached at time t_c occurs when $dD/dt = 0$ in Equation (9.5). Hence, putting $D_c = D$ and $L = L_a 10^{-k_1 t_c}$ it can be shown that

$$D_c = \frac{k_1}{k_2} L_a 10^{-k_1 t_c} \tag{9.7}$$

and occurs at

$$t_c = \frac{1}{k_2 - k_1} \log_{10} \left\{ \frac{k_2}{k_1} \left[1 - \frac{D_a}{L_a} \left(\frac{k_2 - k_1}{k_1} \right) \right] \right\} \tag{9.8}$$

For the conditions in the last example

$$t_c = \frac{1}{0.05} \log_{10} \left\{ \frac{0.20}{0.15} \left[1 - \frac{2.4}{10} \left(\frac{0.05}{0.15} \right) \right] \right\}$$

$$= 1.77 \text{ days}$$

and

$$D_c = \frac{0.15}{0.20} \times 10 \times 10^{-0.15 \times 1.77} = 4.07 \text{ mg/l}$$

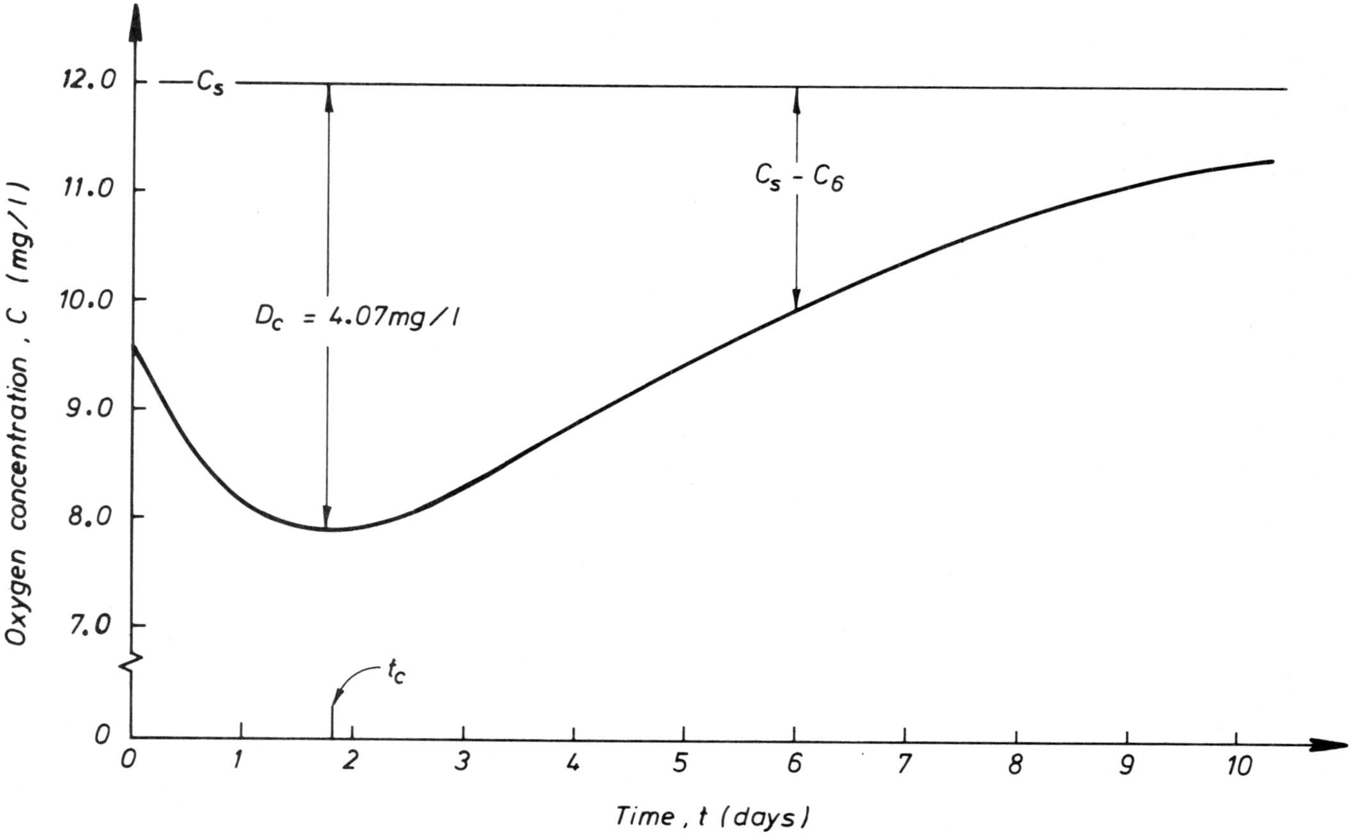

Fig. 9.1 Dissolved oxygen sag curve

The fact that dissolved oxygen and BOD are conservative elements which may be treated like mass for the purposes of calculating dilution of effluent streams, and the conversion of BOD_5 to BOD_{ult} have been dealt with in Chapter 1, Section 1.5.1. The example below is a simple calculation which combines these processes.

Example 9.4

A town discharges 0.05 m^3/s of settled sewage with a BOD_5 of 150 mg/l (for which the reaction rate is 0.17 per day at 20 °C) into a river which has a minimum flow of 0.3 m^3/s. The temperatures of the river and effluent are 15 °C and k_1 and k_2 for the river are found to be 0.2 per day and 0.3 per day by experiment. The dissolved oxygen contents of the effluent and river are 2.1 mg O_2/l and 9.4 mg/l, respectively, prior to mixing and the BOD ultimate of the river is 2 mg/l. Calculate the magnitude of the critical dissolved oxygen deficit and the time at which it occurs.

Calculation

For the sewage BOD_5 test

$$L_0(20\,°\mathrm{C}) = \frac{\mathrm{BOD}_5}{1 - 10^{-0.17 \times 5}} = \frac{150}{1 - 10^{-0.85}}$$

$$= 175 \text{ mg/l}$$

Correcting L_0 to 15 °C gives

$$L_0(15\,°\mathrm{C}) = 175[1 + 0.02(15 - 20)]$$

$$= 158 \text{ mg/l}$$

For the stream dilution of BOD

$$158 \times 0.05 + 2.0 \times 0.30 = L_a(0.05 + 0.30)$$

Thus $L_a = 24.3$ mg/l.

For the stream dilution of dissolved oxygen

$$2.1 \times 0.05 + 9.4 \times 0.3 = C \times 0.35$$

$$C = 8.36 \text{ mg/l}$$

From Equation (4.28), the value of C_s at 15 °C for *fresh* water is 10.2 mg/l; thus $D_a = 10.2 - 8.4 = 1.8$ mg/l. Thus

$$t_c = \frac{1}{0.3 - 0.2} \log_{10} \left\{ \frac{0.3}{0.2} \left[1 - \frac{1.8}{24.3} \left(\frac{0.3 - 0.2}{0.2} \right) \right] \right\}$$

$$= 1.60 \text{ days}$$

and

$$D_c = \frac{0.2}{0.3} \cdot 24.3 \times 10^{-0.2 \times 1.6}$$

$$= 7.75 \text{ mg/l}$$

□ □ □

Fair [3] has proposed modifying the Streeter–Phelps equation by using a cipher f to represent k_2/k_1. The time t_c then becomes

$$t_c = \frac{1}{k_1(f-1)} \log\left\{f\left[1 - \frac{D_a}{L_a}(f-1)\right]\right\} \tag{9.9}$$

and

$$D_c = \frac{L_a}{f} 10^{-k_1 t_c} \tag{9.10}$$

If only the magnitude of D_c is required it is unnecessary to determine k_1 since $k_1 t_c$ is determinable directly from Equation (9.9).

Values of f from 0.5 for backwaters, through 1.5–2.0 for large slow moving rivers to greater than 4 for swift streams are quoted [3].

9.4.1 Determination of k_1 and k_2 in rivers

The value of k_1 found in the stream differs from the values of k_1 found from the BOD tests not only because of the effects of temperature but also because turbulence, dilution, predation and UV disinfection of bacteria influence the oxidation rate. If the ultimate BODs for two stations 1 (upstream) and 2 (downstream) are known, then k_1 can be assessed from

$$k_1 = \frac{1}{\Delta t} \log\left(\frac{L_{a(1)}}{L_{a(2)}}\right) \tag{9.11}$$

where $L_{a(1)}$ and $L_{a(2)}$ are the ultimate BODs at station 1 (upstream) and 2 (downstream) and Δt is the time of flow between stations.

For the determination of k_2, Churchill [4] has suggested that the k_2 (20 °C) value is a function of the mean forward velocity (flow/mean cross-sectional area) V, m/s, and the mean depth (mean cross-sectional area/mean width) R, m, i.e.

$$k_2 \text{ (20 °C)} = 2.18 \frac{V^{0.969}}{R^{1.673}} \tag{9.12}$$

This expression cannot be used where variations in V and R between sample stations become significant, or where stratification of the river occurs. An example of the use of these equations is given below.

Example 9.5

Table 9.1 gives the results of an investigation into a reach of a river. From the information provided, estimate whether a sag curve occurs in the reach, and, if so, determine its position and magnitude.

Calculation

For the determination of k_1

$$L_A = 18.4 \text{ mg/l}; \quad L_B = 5.3 \text{ mg/l}$$

$$V_{mean} = 3.9/13.1 = 0.30 \text{ m/s}$$

thus

$$\Delta t = \frac{65 \times 10^3}{0.3 \times 8.64 \times 10^4} = 2.5 \text{ days}$$

Thus

$$k_1 = \frac{1}{2.5} \log_{10} \frac{18.4}{5.3} = 0.22 \text{ per day}$$

For k_2

$$R_{mean} = \frac{13.1}{9.3} = 1.41 \text{ m}$$

thus

$$k_2(20\,°\text{C}) = 2.18 \frac{0.30^{0.969}}{1.41^{1.673}}$$

$$= 0.38 \text{ per day}$$

Also

$$k_2(10\,°\text{C}) = 0.38 \times 1.024^{(-10)}$$

$$= 0.30 \text{ per day}$$

From Equation (4.28), C_s at 10 °C is 11.4 mg/l, thus $D_a = 11.4 - 6.3 = 5.1$ mg/l.

If a critical deficit occurs, it will occur at time t_c where

$$t_c = \frac{1}{0.30 - 0.22} \log_{10} \left(\frac{0.30}{0.22} \left[1 - \frac{5.1}{18.4} \frac{(0.30 - 0.22)}{0.22} \right] \right)$$

$$= 1.11 \text{ days}$$

i.e. at a distance $0.3 \times (8.64/10^3) \times 10^4 \times 1.11$ km or 28.8 km downstream of A

Table 9.1 River parameters

	Distance (km)	Mean flow (m^3/s)	Mean cross-section (m^2)	Mean width (m)	Ultimate BOD (mg/l)	DO (mg/l)	Temperature (°C)
Station A	0.0	3.9	13.1	9.3	18.4	6.3	10 °C
Station B	65.0				5.3		

the magnitude will be

$$D_c = \frac{0.22}{0.30} \times 18.4 \times 10^{-1.11 \times 0.22}$$

$$= 7.7 \text{ mg/l}$$

Questions

Question 9.1

A BOD test (5 days, 20 °C) gave a result of 257 mg/l with a k_1 value of 0.15 days. Calculate k_1 and the ultimate BOD (L_0) at 10 °C.

Question 9.2

A 20 °C BOD test was carried out by a manumetric method. The readings of the BOD_t were:

BOD_t (mg/l)	62	111	205	270	291	297	299	300	300
Time (days)	1	2	5	10	15	20	25	50	100

Plot $\log_{10}(L_t/L_0)$ against t in days and hence find k_1 (given $L_0 - BOD_t = L_t$ from Section 1.5.1).

Question 9.3

A river contains 10 mg O_2/l and a BOD_{ult} of 2 mg/l and flows at 0.1 m^3/s. It receives a sewage effluent containing 1 mg O_2/l and a BOD_{ult} of 20 mg/l flowing at 0.01 m^3/s. Calculate the size of the critical deficit if the Fair constant f is 1.75. C_s for the river is 11.0 mg/l.

References

1 Discharge of Sewage from Sea Outfalls (1975), in *Supplement to Progress in Water Technology*, Pergamon, Oxford.
2 Streeter, H. W. and Phelphs, E. B. (1925), A study of the Purification of the Ohio River, *US Public Health Service Bull.*, No. 146, Washington DC.
3 Fair, G. M. (1939), *Sewage Wks. J.*, **11**, 445–61.
4 Churchill, M. A. and Buckingham, R. A. (1956), *Sewage Ind. Wastes.*, **28**, 517–37.

Further reading

Nemerow, N. L. (1974), *Scientific Stream Pollution Analysis*, McGraw-Hill, New York.

Appendix 1

Non SI relationships

Many excellent books and older papers contain non metric relationships which need to be transformed into SI units. There are many possible approaches, but the following is suggested as a quick and simple method.

Dimensionally correct equations

Dimensionally correct relationships which have dimensionless constants do not require any transformation *provided coherent* units are used.

Example
The D'Arcy–Weisbach relationship relating head loss (h) to pipe length (l), pipe diameter (d), velocity of flow (v), acceleration due to gravity (g) and dimensionless coefficient f is

$$h = f\frac{v^2 l}{2gd} \text{ or } \frac{f}{2} = \frac{h}{l}\frac{gd}{v^2}$$

Dimensionally

$$\frac{L}{L} \cdot \frac{L}{T^2} \times \mathrm{L}\left(\frac{\mathrm{T}}{\mathrm{L}}\right)^2 = \{\mathrm{LMT}\}^0$$

Choosing any multiples of LM and T would have no affect on the constant, provided all parameters are expressed in terms of similar multiples.

□ □ □

Empirical relationships

Many relationships contain dimensional constants which change with the choice of units.

Example
The Bilham formula (see Chapter 1, Section 1.2) in Imperial units is:

$$N = 1.25 + (w + 0.1)^{-3.55}$$

where N is the number of storms of duration t (h) per 10 years and w is the total rainfall in inches. Since w is usually quoted in millimetres, the expression is unchanged by replacing w (in) by w (mm)/25.4 thus

$$N = 1.25t\,(0.0394w + 0.1)^{-3.55}$$

Seemingly complex examples are also easily converted.

□□□

Example

The length of side weir l (in feet) required to discharge a flow with mean velocity v (feet/s) in a rectangular channel of width b (feet) for an incoming head above the weir of Y_2 (feet) and outgoing head above the weir Y_1 (feet) is given by the Coleman relationship

$$l = 0.548\, bvY_2^{0.13}\left(\frac{1}{\sqrt{Y_1}} - \frac{1}{\sqrt{Y_2}}\right)$$

Multiplying those units in feet by 3.2808 enables the new values to be inserted directly in metres since 1 m = 3.2808ft. Thus

$$l\,3.2808 = 0.548(b\,3.2808)\,(v\,3.2808)\,(Y_2 \times 3.2808)^{0.13}$$

$$\times\left(\frac{1}{\sqrt{(Y_1 \times 3.2808)}} - \frac{1}{\sqrt{(Y_2 \times 3.2808)}}\right)$$

or

$$l = 0.548\, bvY_2^{0.13}\,\frac{(3.2808)^{2.13}}{(3.2808)\sqrt{3.2808}}\left(\frac{1}{\sqrt{Y_1}} - \frac{1}{\sqrt{Y_2}}\right)$$

i.e.

$$l = 1.158\, bvY_2^{0.13}\left(\frac{1}{\sqrt{Y_1}} - \frac{1}{\sqrt{Y_2}}\right)$$

□□□

The SI version of the equation has quoted the constant 1.158 to 1 in 1158 or 0.09%, whereas the previous constant (0.548) was originally quoted to 1 in 548 or 0.2%. An alternative would be to have quoted the constant (in SI) as 1.16, i.e. 1 in 116 or 0.9%. The decision has consciously been made to retain the maximum accuracy of the original expression by the use of 1.158, however it should be noted that the actual accuracy cannot be better than the original 0.2%.

Caution should be used in two circumstances:

(*a*) where volumes are involved, since there is a considerable difference between the US gallon (3.785 l) and the Imperial gallon (4.546 l) (*b*) where the expression lb is used, since lb weight or lb force has dimensions of MLT^{-2}. Thus to convert a stated pressure of 1 lb/sq. in requires the understanding that it is equal to (1 lb mass x 32.18 ft/s^2)/in^2 or (0.453 kg x 9.81 m/s^2)/6.45 x 10^{-4} m^2. The conversion is therefore 1 lbforce/in^2 = 6897 N/m^2. This is usually stated as 6.897 kPa.

Tables A1.1 and A1.2 give conversion factors for the majority of the common parameters found in wastewater treatment.

Table A1.1 Imperial/SI and US/SI parameters and conversion factors [Courtesy of The *Journal of the Institution of Water Pollution Control*]

Parameter	Imperial unit	SI unit	Conversion factor: Imperial gallon	Conversion factor: US gallon
Rainfall	inches	mm	25.4	
Sewer (diameter)	inches	mm	25.4	
Sewage:				
Flow	gal/min	l/s	0.0758	0.0631
	gal/day	m^3/day	0.004 546	0.003785
	mil gal/day	$10^3 m^3$/day	4.546	3.785
BOD per capita	lb BOD/hd day	kg BOD/hd day	0.453 6	
Pumping rate	gal/min	l/s	0.075 8	0.0631
Preliminary Treatment				
Screenings	ft^3/mil gal	$m^3/10^3 m^3$	0.006 23	0.00748
Grit removal:				
Velocity	ft/s	m/s	0.304 8	
Volume of grit	ft^3/mil gal	$m^3/10^3 m^3$	0.006 23	0.00748
Air	ft^3/min ft	m^3/s m	0.001 55	
Primary Treatment				
Capacity	gal/hd	l/hd	4.546	3.785
Surface loading	gal/ft^2 day	m^3/m^2 day	0.0489	0.0407
Upward-flow rate	ft/h	m/h	0.304 8	
Weir loading	gal/ft day	m^3/m day	0.014 9	0.0124
Scraper speed	ft/min	mm/s	5.08	
Secondary Treatment				
Biological filters:				
Loading:				
Hydraulic	gal/yd^3 day	m^3/m^3 day	0.005 946	0.004951
Superficial	gal/yd^2 day	m^3/m^2 day	0.005 437	0.004527
BOD	lb BOD/yd^3 day	kg BOD/m^3 day	0.593 2	
Irrigation rate	gal/ft^2 h	m^3/m^2 h	0.0489	0.0407
Specific surface area	ft^2/ft^3	m^2/m^3	3.28	
Activated-sludge plant:				
Loading	lb BOD/1000 ft^3 day	kg BOD/m^3 day	0.016	

Sludge loading		g BOD/g activated sludge per day		
Sludge-volume index		ml/g		
Air	ft^3/gal	m^3/m^3	6.23	7.48
	ft^3/lb BOD removed	m^3/kg BOD removed	0.062 4	
	$ft^3/min\ ft^2$ plan area	$m^3/s\ m^2$ plan area	0.005 08	
Air pressure	lb/in^2	bar	0.068 95	
Air compressor rating	ft^3/min	m^3/s	0.000 472	
Power	hp/mil gal day	W/m^3 day	0.164	0.197
Tertiary Treatment				
Sand filters:				
Loading	gal/ft^2 h	m^3/m^2 day	1.176	0.979
Backwashing rate	gal/ft^2 min	m^3/m^2 day	70.46	58.71
Grass plots	mil gal/acre day	m^3/m^2 day	1.123	0.936
Sludge Treatment				
Sludge:				
Volume	mil gal	m^3	4546	3785
Production	gal/hd day	m^3/hday	0.004 546	0.003785
	lb dry solids/hday day	kg dry solids/hday day	0.453 6	
	lb volatile matter/hday day	kg volatile matter/hd day	0.453 6	
Calorific value	Btu/lb	kJ/kg	2.326	
Specific resistance	s^2/g at 500 g/cm^2	m/kg at 49 kPa	9.81×10^3	
Sludge gas:				
Production	ft^3/lb volatile matter destroyed	m^3/kg volatile matter destroyed	0.062 4	
Calorific value	Btu/ft^3	kJ/m^3	37.26	
Sludge dewatering				
Yield	lb dry solids/ft^2 h	kg/m^2 h	4.881	
Sludge digestion:				
Loading	lb volatile matter/ft^3 day	kg/m^3 day	16.03	
Sludge heaters:				
Rating	10^6 Btu/h	MW	0.293 1	
Gas consumption	ft^3/h	m^3/h	0.028 32	
Fuel oil:				
Calorific value	Btu/gal	kJ/l	0.232	0.279
	Btu/lb	kJ/kg	2.326	

Table A1.2 Imperial/metric (SI) conversion factors [courtesy of *J. Inst. Water Pollution Control*]

Multiply imperial unit by figure in multiplier column to obtain metric (SI) equivalent; multiply metric (SI) unit by reciprocal to obtain imperial equivalent.

Imperial unit	Multiplier	Metric unit	Reciprocal
acre	0.404 7	hectare	2.471
Btu	1.055	kJ	0.947 8
Btu/ft^2 h	3.155	J/m^2 s	0.317 0
Btu/ft^3	37.26	kJ/m^3	0.026 8
Btu/gal	0.232 1	kJ/l	4.31
Btu/h	0.293 1	W(J/h)	3.412
Btu/lb	2.326	kJ/kg	0.43
°F	0.555 5 (°F – 32)	°C	(1.8 x °C) + 32
ft	0.304 8	m	3.281
ft/min	5.08	mm/s	0.196 9
ft/s	0.304 8	m/s	3.281
ft^2	0.092 9	m^2	10.76
ft^2/ft^3	3.28	m^2/m^3	0.305
ft^3	0.028 32	m^3	35.31
ft^3/ft^2	0.304 8	m^3/m^2	3.281
ft^3/gal	6.23	m^3/m^3	0.160 5
ft^3/lb	0.062 4	m^3/kg	16.1
ft^3/mil gal	0.006 23	$m^3/10^3 m^3$	160.51
ft^3/min	0.000 472	m^3/s	2119
ft^3/min ft	0.001 55	m^3/s m	645.83
ft^3/min ft^2 plan area	0.005 08	m^3/s m^2 plan area	196.85
ft^3/s (cusec)	0.028 32	m^3/s (cumec)	35.31
gal	4.546	litre	0.220
gal	0.004 546	m^3	220
gal/ft	0.014 9	m^3/m	67.114
gal/ft^2	0.049	m^3/m^2	20.408
gal/ft^2 h	1.176	m^3/m^2 d	0.850
gal/ft^2 min	70.56	m^3/m^2 d	0.014 2
gal/min	0.075 8	l/s	13.192
gal/yd^2	0.005 437	m^3/m^2	183.9
gal/yd^3	0.005 946	m^3/m^3	168.2
hp	0.745 7	kW	1.341
hp/mil gal	0.164	W/m^3	6.098
in	25.40	mm	0.039 37
in^2	645.2	mm^2	0.001 55
in^3	16.39×10^3	mm^3	0.061×10^{-3}
lb	0.453 6	kg	2.205
lb/ft^2	4.881	kg/m^2	0.204 9
lb/ft^3	16.03	kg/m^3	0.062 43
lb/$10^3 ft^3$	0.016	kg/m^3	62.5
lb/in^2	0.068 95	bar	14.503
lb/yd^3	0.593 2	kg/m^3	1.686
mil gal	4546	m^3	0.220×10^{-3}
mil gal/acre	1.123	m^3/m^2	0.890
mile	1.609	km	0.621 4
ton	1.016	tonne	0.984 2
yd	0.914 4	m	1.094
yd^2	0.836 1	m^2	1.196
yd^3	0.764 6	m^3	1.308

Appendix 2

Flow in pipes

The nomogram in Fig. A2.1 may be used to determine the interrelationships between the discharge, pipe diameter, gradient and velocity for sewer pipes running full. It is reproduced by courtesy of the Christchurch Drainage Board, New Zealand.

It is based on the Colebrook–White equation and incorporates the D'Arcy–Weisbach relationship (see Section 1.3.3). Note that it will not produce identical results to the Manning formula when $n = 0.012$, e.g. a straight edge aligned on the nomogram at $D = 300$ mm and $S = 10$ per thousand (0.01) gives a velocity of 1.4 m/s whereas the Manning formula for a pipe running full ($n = 0.012$) would give a velocity of 1.48 m/s.

Use of the Moody or universal flow diagram

From Section 1.3.3, the frictional head loss h_f in a pipe of length L(m) and diameter D(m) is related to the velocity of flow V(m/s) and the gravitational acceleration g(m/s^2) by a dimensionless friction factor f in the D'Arcy–Weisbach equation:

$$\frac{h_f}{L} = f\frac{V^2}{2gD}$$

The value of f varies with the dimensionless Reynolds number R_e according to Colebrook–White formula

$$\frac{1}{\sqrt{f}} = -2\log_{10}\left(\frac{\epsilon}{3.7D} + \frac{2.51}{R_e\sqrt{f}}\right)$$

where $R_e = (VD)/\upsilon$, υ being the kinematic viscosity of the fluid and ϵ the equivalent sand grain roughness of the pipe in metres (that pipe roughness producing the same frictional head loss that would be produced by coating the inside of a smooth pipe with sand of diameter ϵ). If a value for ϵ is assumed for a pipe, the use of the above equations involves iterative calculations to find h_f. However, the universal pipe friction diagram or Moody[1] diagram (Fig. A2.2) greatly eases the calculations.

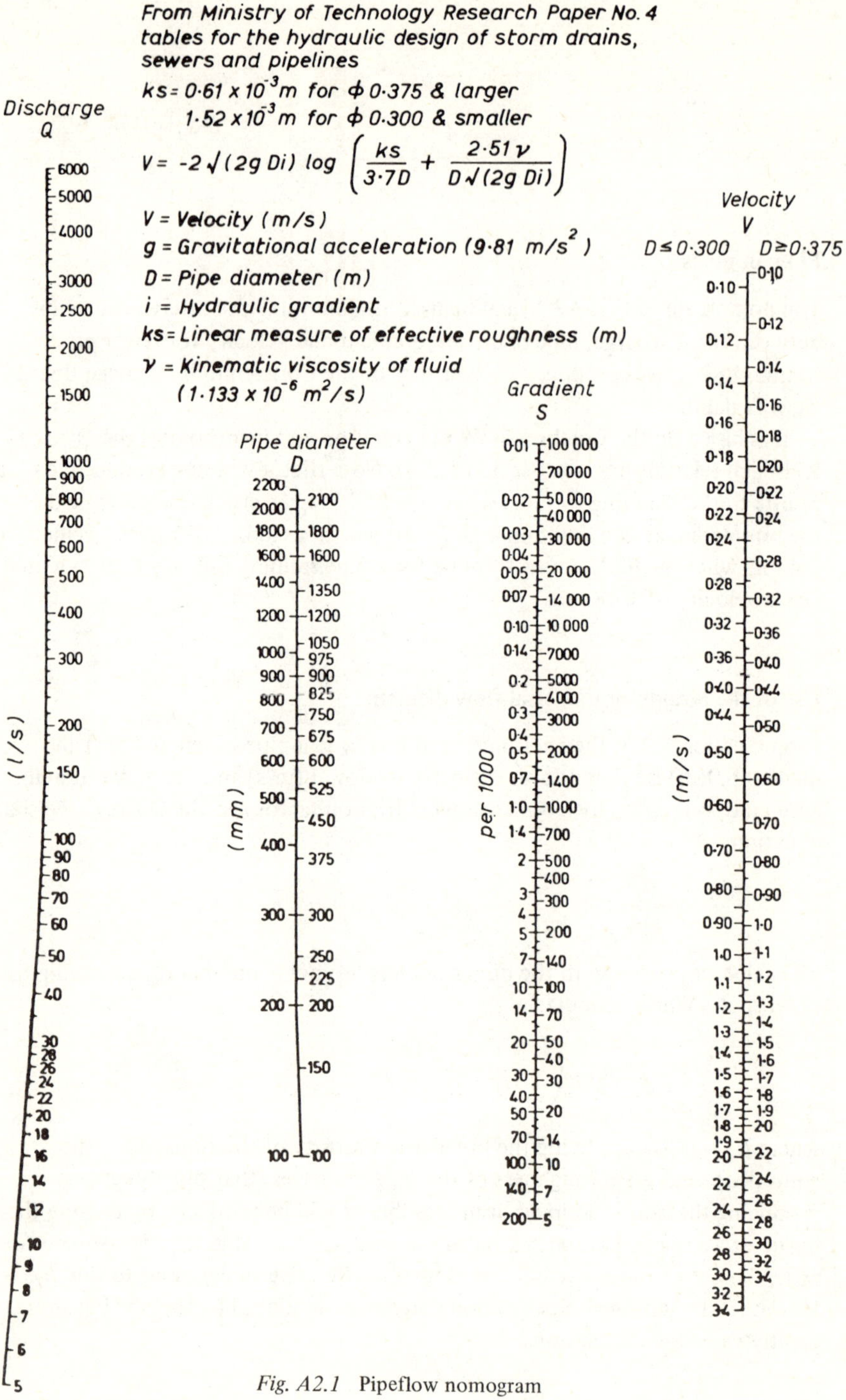

Fig. A2.1 Pipeflow nomogram

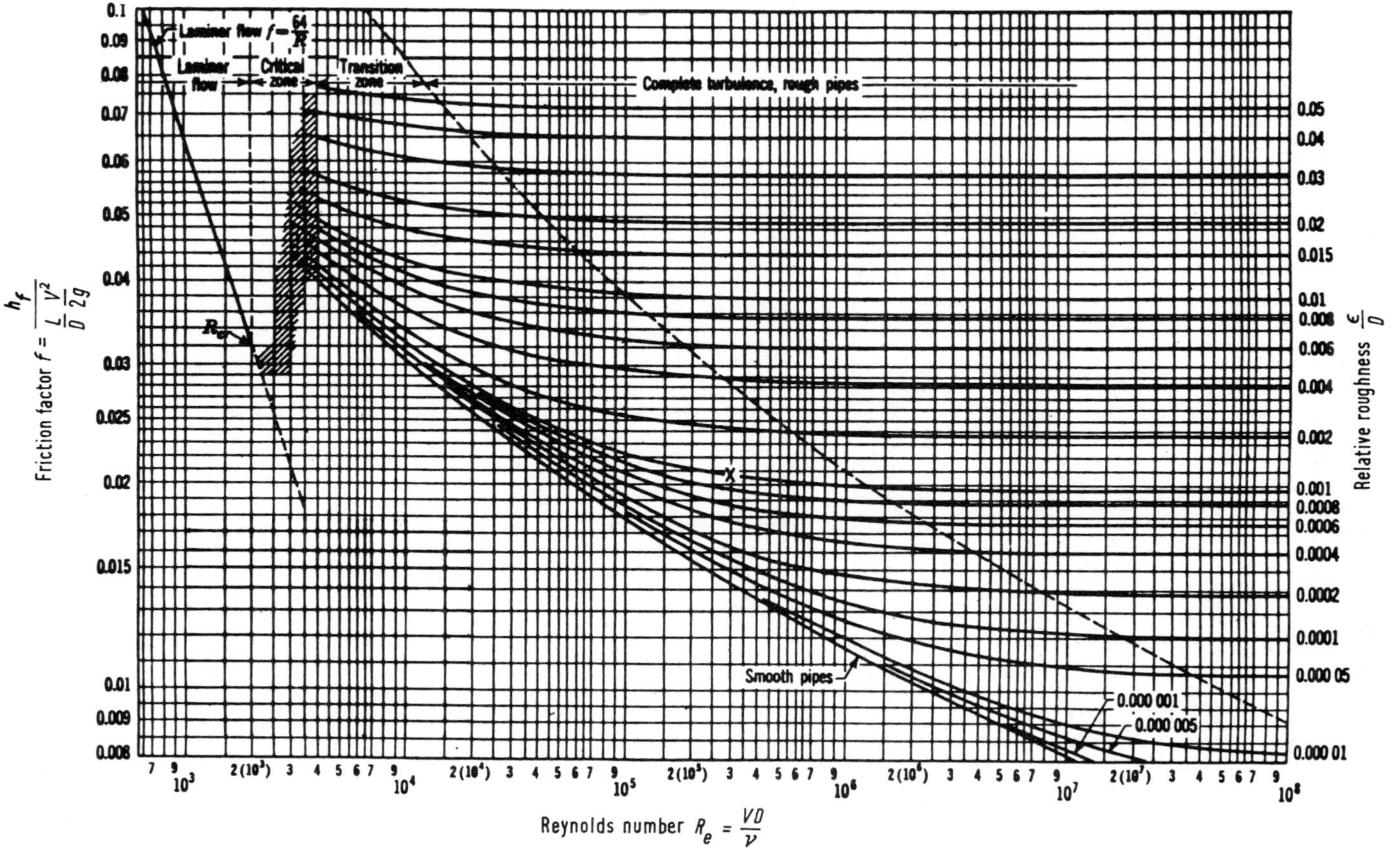

Fig. A2.2 Universal or Moody diagram (After Moody[1]), by courtesy of the American Society of Mechanical Engineers).

Use of the universal diagram

Assume a value for ϵ of 0.002 m for a 100 m long pipe of 0.200 m inside diameter. ϵ/D then is 0.001. If it is desired to calculate h_f for a maximum flow of 60 l/s then the pipe velocity is

$$V = \frac{6 \times 10^{-2}}{\pi(0.2)^2/4} \text{ m/s or } 1.91 \text{ m/s}$$

At 20 °C, $\upsilon = 1.3 \times 10^{-6}$ m^2/s and thus

$$R_e = \frac{1.91 \times 0.200}{1.3 \times 10^{-6}} = 2.94 \times 10^5 \cong 3 \times 10^5$$

Following the *curve* for $\epsilon/D = 0.001$ from the right hand side (ordinate) of the diagram to the junction with the vertical line at $R_e = 3 \times 10^5$ (on the abcissa) a point x is found which, reading on the left hand side (ordinate) of the graph, is equivalent to a value of f of approximately 0.021.

Substituting this value in the D'Arcy–Weisbach equation gives a value for h_f of

$$h_f = 0.021 \times \frac{1000 \times (1.91)^2}{2 \times 9.81 \times 0.2} = 19.5 \text{ m}$$

Reference

1 Moody, L. F. (1944), *Trans. ASME,* **66.**

Appendix 3

Biological systems

Many micro-organisms and some larger organisms are commonly used in wastewater treatment processes. Many secondary biological processes, e.g. activated sludge or percolating filters, are engineering systems designed to speed up processes of waste stabilization (oxidation) commonly found in rivers and lakes. The following material is intended to give basic information on the organism types involved. The most important organisms in wastewater treatment are as follows.

Rotifers, worms and larvae

Rotifers are multicellular higher organisms which have rotating cillia or hairs which appear to push or waft food particles into their sac-like bodies. Rotifers graze the less complex organisms, but as they are a more complex animal, they require a stable, well aerated environment.

Worms and larvae are the most significant of the remaining higher organisms. They are important in percolating filters, as they have an influence in maintaining the void spaces between stones. The worms and larvae exist in the upper layers of a percolating filter, where they graze other organisms. Their movement deeper into the bed during the winter can adversely affect the purification efficiency of the filter.

Fungi, protozoa and algae

These range from less than 10 μm to greater than 10^3 μm. Fungi are multicellular, non photosynthetic plants which use organic matter as a food source. They metabolize similar materials to that of bacteria but require less nitrogen and can grow at lower pH values. Fungi are larger than bacteria and are filamentous; they have poorer settling properties than bacteria and it is therefore usual to discourage the formation of fungi in sewage treatment processes.

Protozoa are single celled animals which can utilize solid substances and reproduce by simple division. This group includes types which are motile, e.g. free swiming ciliates, and non-motile forms, e.g. stalked ciliates. Protozoa can utilize bacteria as a food source, so are important in sewage treatment processes, as they help to maintain a balanced population. Protozoa become more predominant in waters which contain less organic pollution. Thus, in any sequential sewage treatment process, the first stages rely almost entirely upon bacteria, while in later stages protozoa become more evident.

Algae are unicellular or multicellular plants which like protozoa and bacteria respirate or require oxygen for their life processes. Most algae can also photosynthesize new cells from CO_2 and nutrients under the influence of sunlight. The combination of bacterial oxidation of organics to give CO_2, and the algal photosynthetic production of oxygen is the basis of the facultative oxidation pond process.

Bacteria

Bacteria range from 10^{-1} μm to 5 μm and may be grouped as below.

A *Autotrophic* utilizing carbon from CO_3^{2-}; HCO_3^- or CO_2 in solution.
- A.1 Chemosynthetic. In addition to C they use an inorganic substance. e.g. Nitrosomonas oxidize ammonia to nitrite ion.
- A.2 Photosynthetic. Not generally important in wastewater treatment.

B *Heteratrophic* utilizing more complex soluble substrates containing carbon.
- B.1 Nitrogen fixers. Important in agriculture.
- B.2 Non-exacting organisms. These can utilize carbon from a wide range of substrates, e.g. the bacteria used as an indicator organism *E. coli.*
- B.3 Exacting organisms. Require specific substances often particular amino acids.

The conditions under which bacteria grow best are given below under different headings:

Nutrients

In order to synthesize new proteinaceous material bacteria require C, H, O, N, P, S with traces of a large number of anions and cations. Ratios of BOD_5 to N and P of 20 and 100 (or less), respectively, are usually satisfactory and it is rare for other elements to be limiting. For domestic sewage N, P, and S concentrations of 30–50 mg/l, 22 mg/l and 22 mg/l, respectively, have been quoted for settled sewage. It is therefore only usual for nutrient imbalance to limit bacterial growth in wastewaters containing industrial wastes.

pH

Bacterial reactions are achieved by using proteinaceous enzymes. These consist of a basic structure of amino acids whose actions are considerably influenced by the acidity of the solution. The enzyme activity (ordinate) plotted against pH (abcissa) often produces a bell shaped curve with steep sides whose upper plateau spans perhaps 2–3 pH units. Thus in wastewater treatment, reactions often cease if the pH drops below pH 5 or goes above pH 10, and optimum conditions often are restricted to a pH range from 6.5 to 8.5.

Temperature

Although some bacteria can exist (sometimes as a spore) at extreme temperatures, three general types are found as shown in Table A3.1.

Table A3.1 Bacterial types according to temperature range

Bacterial type	Optimum growth range (°C)	Viable growth range (°C)
Psychrophiles	15–20	0–30
Mesophiles	25–40	5–45
Thermophiles	50–55	25–80

Oxygen concentration

Wastewater oxidative bacteria and most purifying protozoa require *aerobic* conditions which is usually taken as dissolved oxygen (DO) levels in excess of 1 mg/l in secondary biological treatment stages. Nitrifying bacteria work best at DO levels $>$ 2 mg/l. Some bacteria, e.g. denitrifying bacteria, operate best when the solution is almost completely devoid of oxygen, i.e. when the oxidation–reduction potential is mildly reducing. The *anoxic* conditions are important in a nitrogen stripping process (see Chapter 5, Section 5.4.4).

Other bacteria which carry out reactions of a strongly reducing type, e.g. mesophilic methane fermenters, require anaerobic conditions to occur.

Certain bacterial types, e.g. those responsible for acid formation in mesophilic digesters, may grow under both aerobic and anaerobic conditions and are referred to as facultative organisms.

Further reading

Barnes, D. and Wilson, F. (1978), Biological background to wastewater treatment, in *Chemistry and Unit Operations in Sewage Treatment*, Applied Science Publishers, London.

Mara, D. D. (1974), *Bacteriology for Sanitary Engineers*, Longmans, Edinburgh.

Hugo, W. B. (1964), *An Introduction to Microbiology*, Heinemann, London.

Appendix 4

Secondary biological treatment processes – typical loading rates

Activated sludge

	SLR or F/M ratio (kg BOD/kg MLSS day)	Oxygen requirement (kg O_2/kg BOD removed)	Aeration volumetric loading (kg BOD/m^3 day)	Aeration volume/capita (m^3) based on BOD load to secondary process kg/d in brackets	Sludge wastage rate (kg/kg BOD removed)
Extended aeration	0.05–0.10	2.0–2.5	0.15–0.30	0.200–0.250 (0.06)	0.4–0.6
Conventional process	0.25–0.35	1.1–1.3	0.5–0.65	0.05–0.08 (0.04)	0.5–0.8
High rate process	0.5–0.6	0.6–0.8	1.3–2.0	0.02–0.03 (0.04)	0.8–1.0

Percolating Filter

	BOD loadings (kg BOD/m^3 media . day)	Hydraulic loading (m^3/m^3 media . day) at maximum flow)	Sludge yield (kg/kg BOD removed)	Specific surface area (m^2/m^3)
Single pass filter	0.1–0.2*	0.3–0.8*	0.4–0.6	40
Alternating double filtration	0.2–0.4*	0.8–1.2*	0.4–0.6	40
High rate (plastic media) filter	up to 6.0	up to 20	0.8–1.0	>60

* These rates are to give an effluent BOD of 30 mg/l.

Appendix 5

Dissolved oxygen saturation values

The following table shows dissolved oxygen saturation values in saline waters in equilibrium with water saturated air at normal pressure (101.3 kN/m^2) (after Whipple, G. C. and Whipple, M. C. (1911), *J. Amer. Chem. Soc.*, **33**, 362–65).

Temperature (°C)	Chloride ion concentration (p.p.m.) 0	5000	10000	15000	20000	Difference (/100 p.p.m.) Cl^-
	Dissolved oxygen (mg/l)					
0	14.6	13.8	13.0	12.1	11.3	0.017
1	14.2	13.4	12.6	11.8	11.0	0.016
2	13.8	13.1	12.3	11.5	10.8	0.015
3	13.5	12.7	12.0	11.2	10.5	0.015
4	13.1	12.4	11.7	11.0	10.3	0.014
5	12.8	12.1	11.4	10.7	10.0	0.014
6	12.5	11.8	11.1	10.5	9.8	0.014
7	12.2	11.5	10.9	10.2	9.6	0.013
8	11.9	11.2	10.6	10.0	9.4	0.013
9	11.6	11.0	10.4	9.8	9.2	0.012
10	11.3	10.7	10.1	9.6	9.0	0.012
11	11.1	10.5	9.9	9.4	8.8	0.011
12	10.8	10.3	9.7	9.2	8.6	0.011
13	10.6	10.1	9.5	9.0	8.5	0.011
14	10.4	9.9	9.3	8.8	8.3	0.010
15	10.2	9.7	9.1	8.6	8.1	0.010
16	10.2	9.5	9.0	8.5	8.0	0.010
17	9.7	9.3	8.8	8.3	7.8	0.010
18	9.5	9.1	8.6	8.2	7.7	0.009
19	9.4	8.9	8.5	8.0	7.6	0.009
20	9.2	8.7	8.3	7.9	7.4	0.009
21	9.0	8.6	8.1	7.7	7.3	0.009
22	8.8	8.4	8.0	7.6	7.1	0.008
23	8.7	8.3	7.9	7.4	7.0	0.008
24	8.5	8.1	7.7	7.3	6.9	0.008
25	8.4	8.0	7.6	7.2	6.7	0.008
26	8.2	7.8	7.4	7.0	6.6	0.008
27	8.1	7.7	7.3	6.9	6.5	0.008
28	7.9	7.5	7.1	6.8	6.4	0.008
29	7.8	7.4	7.0	6.6	6.3	0.008
30	7.6	7.3	6.9	6.5	6.1	0.008

Appendix 6

Information acquisition

The following search procedures are listed in terms of the information required and the available information. The source material is that most likely to be helpful.

(a) Analytical data pertaining to wastewater treatment is required – no reference available. Example of data required: composition of tannery effluent; seasonal nutrient variation in lakes or % removals of SS by multi media filters.

1 (suggested source) Bond, R. G. (Ed) (1974), *Handbook of Environmental Control, Vol. IV, Wastewater Treatment and Disposal*, Chemical Rubber Co. Press, Ohio, USA.

Comment: Reference 1 is a comprehensive abstraction of information pertaining to wastewater treatment given in table form.

(b) Process design data required – no previous reference. Example: loading rate of percolating filters; infiltration rate into sewers.

2 (Source: USA practice) Amer. Soc. Civil Engrs. (1977), *Wastewater Treatment Plant Design*, Water Pollution Control Fedn. Washington DC.

3 (1973), *Recommended Standards for Sewage Works*. Report of Committee of the Great Lakes – Upper Mississippi River Board of State Sanitary Engineers, Health Education Service, Albany NY.

4 (UK practice) *Manuals of British Practice in Water Pollution Control*, Published by Journal of the Institute of Water Pollution Control, London.

5 *British Standards Yearbook*, British Standards Institution, HMSO, London (relevant standards listed, e.g. CP 2005: 1968, Sewerage).

6 (European) Degrémont, G. (regularly updated), *Water Treatment Handbook* (English version). Société Générale d'Epuration et d'Assainissement, Rueil Malmaison, France.

Comment: Reference 6 is mistitled, it now contains much of interest to wastewater engineers and scientists.

(c) Information required on a specific subject – no reference, e.g. sulphide in sewers or nitrification in the activated sludge process.

7 (Source) (1973), *Water pollution abstracts*, Ministry of Technology, HMSO, London.

8 (1974), *Water Research Centre Information*, WRC, Medmenham, Bucks.

Comment: References 7 and 8 contain abstracts culled from a wide range of sources. Microfiche index for 1974–78 available for Reference 8.

(d) Particular publication required – name of one author available (approximate date is known)

Source: Name index in 7 and 8 or

9 *Engineering Index Annual*, Engineering Index Inc., NY (published yearly in sections).

Comment: Reference 9 is a powerful source of publications in the general engineering fields.

(e) Subject information requires updating, a reference post 1961 is available.

10 (Source) *Science Citation Index*, Philadelphia Institute for Scientific Information, PA (yearly publication).

Comment: The index lists the papers published within that year which have quoted a *previous* article. Provided a main stream paper is available, it enables updating of subject bibliographies to be carried out.

In addition to the above, commercial, computerized search facilities are available. Advertisements are carried in the literature.

Answers

Chapter 1

1.1 116 m^3/h
1.2 27.0 m

Chapter 2

2.1 1.37 m and 0.95 m
2.2 1.9 mm

Chapter 3

3.1 For $Q = 0.02\ m^3/s$
$1 - x_0 \cong 0.43$
$x_T \cong 0.59$ or 59%

for $Q = 0.08\ m^3/s$
$1 - x_0 \cong 0.34$
$x_T = 0.41$ or 41%

3.2 0.5 mm diameter, $V = 81$ mm/s
2 mm diameter, $V = 0.28$ mm/s
3.3 $\psi_L \cong 3.5\ kg/m^2\ h$
Area $\cong 309\ m^2$
3.4 (*a*) $C_u \cong 8.3\ kg/m^3$
(*b*) $C_u \cong 10.0\ kg/m^3$
(*c*) $C_u \cong 13.9\ kg/m^3$
3.5 $L/D = 8$; $D = 3.13$ m; $L = 25$ m; $L/B_0 = 1250\ m^2$;
$B_0 = 50$ m and $n = 5$ tanks

Chapter 4

4.1 $\theta_c = 6.5$ days; SLR = 0.19 per day; SUR = 0.23 per day
4.2 $K_s = 100$ mg/l; $t_d = 0.87$ h
4.3 11.3 days
4.4 4 stages, approximately 140 discs (see Reference 8)
4.5 969 m^3; 29 mg/l

Chapter 5

5.1 9.3 days
5.2 27.2 kg Cl_2/day
5.3 28 mg/l

Chapter 6

6.1 $r = 2.85 \times 10^{12}$ m/kg
6.2 Originally total heat required = 3.01×10^7 kJ/day
Gas produces 2.08×10^7 kJ/day
Suggestion (*a*) total heat required = 2.73×10^7 kJ/day
gas produces 1.80×10^7 kJ/day
Suggestion (*b*) total heat required = 1.20×10^7 kJ/day
gas produces 2.08×10^7 kJ/day
6.3 SC = 24%
6.4 2.1×10^{-3} kg/m^{2}h ($f_c = 1.13$, $C_1 = 20.4$ and $A_f = 0.205$)

Chapter 7

7.1 $Q \cong 133$ m^3/h; $P = 12.2$ kW
7.2 3 l/s

Chapter 9

9.1 $k_1 = 0.095$; $L_0 = 250$ mg/l
9.2 $k_1 = 0.1$ (note that if $BOD_t = 300$, $\log_{10}(L_t/L_0)$ is infinity; only the period up to 25 days can be plotted)
9.3 1.85 mg/l

Index

The main references are indicated by a bold page number